NUCLEAR ARCHIPELAGO

Elizabeth Tynan is a professor in the Graduate Research School at James Cook University, Townsville. She is a prominent writer and researcher of the era of British atomic testing in Australia. Her book *Atomic Thunder: The Maralinga story* (NewSouth, 2016) won the Prime Minister's Literary Award (Australian History) and the CHASS Australia Book Prize in 2017. Her follow-up book *The Secret of Emu Field: Britain's forgotten atomic tests in Australia* was published by NewSouth in 2022.

'In this meticulously researched and beautifully written account of the British nuclear tests conducted off the coast of Western Australia in the 1950s, Elizabeth Tynan provides an excoriating examination of the nuclear politics of the time. As the final book in her trilogy covering the atomic testing program allowed by the Australian government yet conducted in virtual secrecy by Britain, *Nuclear Archipelago* closes the circle on one of the most shameful episodes of sovereign abasement in modern Australia: atomic colonialism, living history, shameful cover-up and more. This book contains pungent lessons for today and provides a heart-wrenching legacy for all those innocently affected by these callous acts – a definite must-read.'

PETER GARRETT, lead singer of Midnight Oil and former federal minister

'Through sensitive scholarship and powerful story-telling, Elizabeth Tynan juxtaposes an awesome *new* energy against the casual *old* injustice of a global empire's desperation to keep its fading influence. *Nuclear Archipelago* is a compelling account of Britain's astonishing indifference to the land, people and ecology of Australia, and of our own governments' feckless complicity. In exposing these terrible, still lingering events, the third book in Tynan's trilogy is at once vital history, and a call for justice. It should also make you angry.'

PROFESSOR MARK KENNY, Director of the Australian Studies Institute, Australian National University

'This multi-volume work must rate as one of the most significant projects currently being undertaken in the study of Australian history.'

SIMON CATERSON, *The Australian*

NUCLEAR ARCHIPELAGO

SECRETS, POWER AND THE BIGGEST ATOMIC BLAST IN AUSTRALIA

ELIZABETH TYNAN

NEWSOUTH

UNSW Press acknowledges the Bidjigal people, the Traditional Owners of the unceded territory on which the Randwick and Kensington campuses of UNSW are situated, and recognises the continuing connection to Country and culture. We pay our respects to Bidjigal Elders past and present.

A NewSouth book

Published by
NewSouth Publishing
University of New South Wales Press Ltd
University of New South Wales
Sydney NSW 2052
AUSTRALIA
https://unsw.press/

Our authorised representative in the EU for product safety is Mare Nostrum Group B.V., Doelen 72, 4831 GR Breda, The Netherlands (gpsr@mare-nostrum.co.uk).

First published 2026

NATIONAL LIBRARY OF AUSTRALIA
A catalogue record for this book is available from the National Library of Australia

ISBN 9781761170218 (paperback)
9781761179440 (ebook)
9781761178719 (ePDF)

Cover design Mika Tabata
Cover images iStock/1129017973 and Shutterstock/2256598363
Internal design Josephine Pajor-Markus
Printer Griffin Press

UNSW
SYDNEY

Frank Rosevere Burden
08/01/1930 – 20/07/2025

In loving memory

CONTENTS

FOREWORD

DR CARMEN LAWRENCE AO, FORMER PREMIER OF WESTERN AUSTRALIA

Elizabeth Tynan's meticulous excavation of nuclear testing on the Monte Bello Islands in the 1950s shows clearly why the study of history is of vital importance in dangerous times. Her dissection of the scientific, political and bureaucratic manoeuvres surrounding the decisions to undertake the tests on these remote – and stunningly beautiful – Australian islands reveals mindsets that are still disturbingly evident: secrecy and dissimulation about national security, carelessness about people's health and the environment, refusal to accept responsibility for past mistakes, and a startling indifference to the threat posed by nuclear weapons.

In its search for relevance in the Cold War world, Britain, its power and influence in decline, but its arrogance towards the former colonies intact, tested nuclear weapons on another country's land. In doing so, and with the subservient assistance of the Australian government, it deliberately exposed its own defence and scientific personnel, as well as the Australians working beside them, to unknown quantities of radiation as they unleashed a fundamental force of the universe. As Tynan makes clear, 'Britain has never properly reckoned with its nuclear test activities. To this day, secrecy reigns'. Wrongs have not been acknowledged, much less righted, despite a scathing Royal Commission report and persistent advocacy for those who suffered long-term damage from radiation and whose lives were upended by the tests.

Among the questions raised by nuclear testing is what responsibility scientists should take for the consequences of their work – with knowledge comes responsibility. Nuclear physicist Joseph Rotblat was clear that this question should not be avoided – although it often was in the mad race to acquire nuclear weapons after the Second World War. After the discovery of nuclear fission in the 1930s, Rotblat worked on fission in the UK and soon found himself at the centre of the Manhattan Project at Los Alamos, New Mexico. Along with a minority of the other scientists involved, he began to ask questions about the morality of working on a weapon of mass destruction and, on leaving the project, devoted the remainder of his life to alerting the world to the dangers posed by nuclear weapons. Such scruples were clearly not front of mind during the Monte Bello tests. The disorganised scramble surrounding some of the test preparations illustrates a degree of insouciance that is truly startling.

It's impossible to read Elizabeth Tynan's account, threaded through with examples of the hubris and ignorance of many of the players, without remembering what else was launched by these tests and others like them – the rapid proliferation of nuclear weapons. Despite the best efforts of those who could see what was at stake, the nuclear club expanded through the ensuing decades and the size of nuclear arsenals exploded, giving precise meaning to the threat of mutually assured destruction. After a period in which it had seemed that the major powers were willing to scale back their nuclear programs and reduce the number of weapons, the global non-proliferation system is now on the brink of collapse, according to the International Atomic Energy Agency. Our own Australian government seems untroubled by this trend and, despite promises, has yet to join the majority of the world's nations in signing up to the 2021 Treaty on the Prohibition of Nuclear Weapons.

The momentum for nuclear disarmament has stalled at a time when the possible use of nuclear weapons has been entertained, the 'nuclear taboo' has been strained: Russia has threatened to use nuclear weapons in its war against Ukraine, and an Israeli Cabinet member spoke of the possible use of nuclear arms in Gaza. In its 2025 annual report, the Stockholm International Peace Research Institute said that the United States, Russia, the United Kingdom, France, China, India, Pakistan, North Korea and Israel – countries confirmed or believed to have nuclear weapons – are all in the process of modernising their arsenals. Other countries, such as Iran, have ambitions to join them. The institute warned of a 'dangerous new nuclear arms race ... at a time when arms-control regimes are severely weakened', adding that 'we are closer to nuclear war than we have been for a long time'. There are currently around 14 000 nuclear warheads, 4000 of which are ready to be launched at a moment's notice from missile silos, submarines and aircraft.

As this is happening, we need history to remind ourselves of the carnage wrought by the two bombs that were detonated – at Hiroshima and Nagasaki – especially since the average age of the A-bomb survivors able to bear witness is now 85. One of the survivors, Terumi Tanaka, spoke for the Hibakusha campaign (survivors of the Hiroshima and Nagasaki bombings) as they accepted the Nobel Peace Prize in 2024. He remembered the day the atomic bomb was dropped on his city, Nagasaki. He was 13 at the time and lived about 3 kilometres east of ground zero. Miraculously he survived, but as he ventured out into the ruined city with his mother to find his aunts, he

> saw the full devastation of the bombing ... blackened ruins spread out as far as the port of Nagasaki, some three kilometers away ... all the houses along the path, down to the foot of the mountain, were burnt to the ground and corpses lay scattered around them. Many people who were badly injured or burned, but still alive, were left unattended, with no help whatsoever.

The Nagasaki bomb exploded 600 metres above the city. The energy released caused shock waves that crushed houses and the people in them. Heat rays severely burned people and ignited fires, incinerating people in their homes. Radiation from neutron and gamma rays caused death, both immediately and slowly, as bodies succumbed to radiation. By the end of 1945, the death toll in the two cities was approximately 140 000 in Hiroshima and 70 000 in Nagasaki. Some 400 000 more suffered injuries from exposure to radiation. Since then, countless others have been harmed by the production and testing of atomic weapons, their lands and seas irradiated and their cultures disrupted – Moruroa, Ekker, Semipalatinsk, Maralinga, Bikini Atoll and the Monte Bello Islands.

Today most of the world seems to accept the unimaginable scale of the nuclear risk as a grim, unavoidable reality, rarely stopping to consider the insanity of these weapons. Yet, learning to live with them is a grave mistake – both genocidal and suicidal, as Martin Luther King said. If even a few of today's nuclear weapons were detonated, the debris and smoke from the firestorms shot high into the atmosphere would cool the Earth, darkening and drying the surface, obliterating food crops, starving billions and making our planet virtually uninhabitable for generations. Knowing how we came to accommodate this reality is an important story in all its elements: the science, the technology, the testing and manufacture of the bombs, the politics that continues to tolerate the nuclear arms race that threatens us all. Knowing the story may help us craft better responses to these threats.

MONTE BELLO ISLANDS

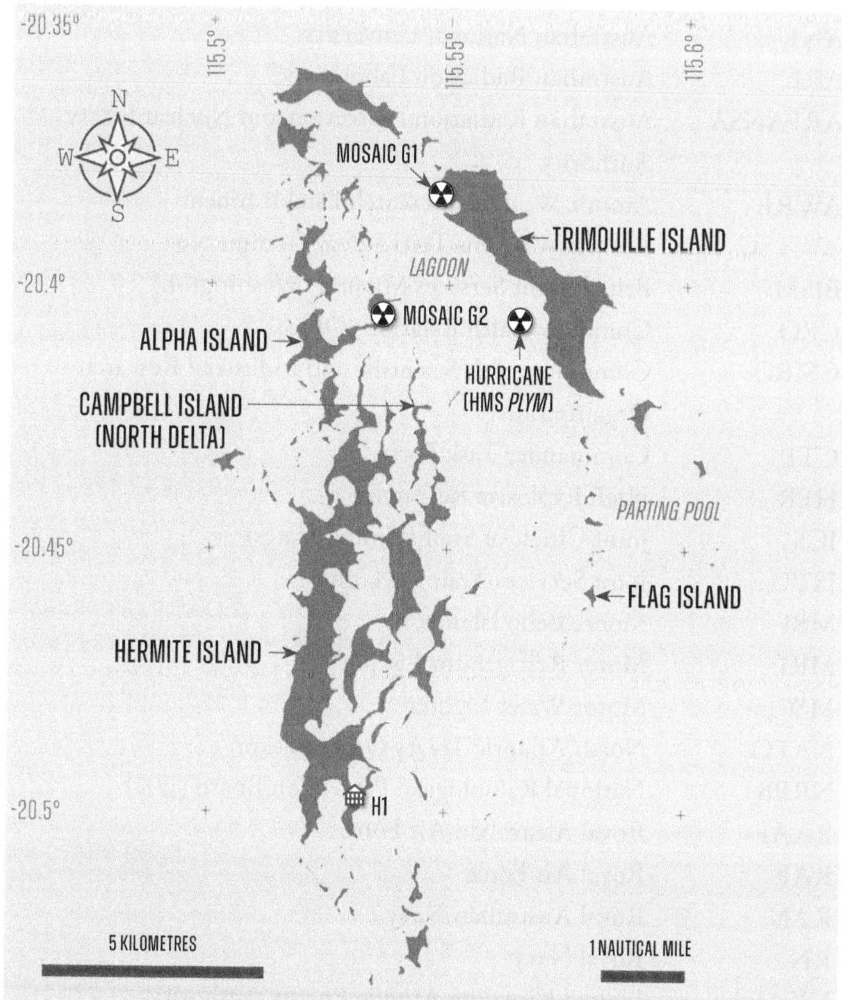

SIMPLIFIED MAP COURTESY OF DR DARREN HOLDEN

ABBREVIATIONS

AERE	Atomic Energy Research Establishment
AIRAC	Australian Ionising Radiation Advisory Council
ANU	Australian National University
ARL	Australian Radiation Laboratory
ARPANSA	Australian Radiation Protection and Nuclear Safety Authority
AWRE	Atomic Weapons Research Establishment
AWTSC	Atomic Weapons Tests Safety Committee
BJSM	British Joint Services Mission (Washington)
CRO	Commonwealth Relations Office
CSIRO	Commonwealth Scientific and Industrial Research Organisation
CTF	Commander Task Force
HER	High Explosive Research
JCS	Joint Chiefs of Staff (United States)
JSTU	Joint Services Training Unit
MBI	Monte Bello Islands
MRL	Motor Refrigerator Lighter
MWL	Motor Water Lighter
NATO	North Atlantic Treaty Organisation
NRPB	National Radiological Protection Board (UK)
RAAF	Royal Australian Air Force
RAF	Royal Air Force
RAN	Royal Australian Navy
RN	Royal Navy
UKAEA	United Kingdom Atomic Energy Authority
UKMOSS(A)	United Kingdom Ministry of Supply Staff (Australia)
UKSLS	United Kingdom Services Liaison Staff

MEASUREMENTS

During the period of the Monte Bello tests, Australia used imperial measurements:

1 inch = 2.54 centimetres
1 foot = 0.309 metres
1 yard = 0.914 metres
1 mile = 1.609 kilometres
1 ton = 0.907 tonne

Author's note: I have chosen in this book to use the rendering of the name of the archipelago that was most common at the time of the tests in the 1950s, and later at the Royal Commission in the 1980s. Most historical documents use the two words. In more recent times, writers and commentators have tended to use the single word, 'Montebello'. This rendering is true to the origins of the name from the Battle of Montebello, but it is not the version that I favour. Therefore, throughout this book I use 'Monte Bello', unless quoting from a source that uses the alternative.

through the late 1940s, once the go-ahead had been given to the British bomb, to overcome the many physics and engineering challenges to make Penney's design a reality. HER changed its name to the Atomic Weapons Research Establishment (AWRE) in 1953.

Joining Penney were Christopher Hinton and John Cockcroft, who together formed the nucleus of Britain's postwar atomic ambitions. Penney would design and test the British bombs, engineer Hinton would create the nuclear fuel for them at the Windscale reactor on the Cumbrian coast of north-west England, while physics Nobel Laureate Cockcroft would run the Atomic Energy Research Establishment (AERE) at Harwell to advance basic understanding of nuclear chemistry, metallurgy, reactor design and radiological monitoring, supplying the skilled workforce for the weapons program. Lindemann constantly assured Churchill that the Americans were superfluous to the British bomb-building effort, and he was able to say this with plausible confidence because of these three men. By this time, the Americans were less concerned with fission-only devices (with explosive power measured in kilotons) as they turned their resources to fusion weapon technology (hydrogen bombs, with explosive power measured in megatons), well ahead of the same policy shift in Britain. In a note to Churchill six weeks after the first Totem test at Emu Field in South Australia in 1953, Lindemann said that discussions with Penney, Cockcroft and Hinton had convinced him that exchange of information with America on the production of fissionable material and the design of bombs was no longer of great importance.

> Six or eight years ago it could have saved us a great deal of money and effort. But today we are convinced that we know practically as much as the Americans do. Whilst we should of course like our scientists and engineers to be able to talk to theirs freely on matters of joint interest, we are convinced that it would not be worth making any sacrifices to achieve this.[20]

By this stage, Britain had exploded three atomic weapons in Australia. Britain had succeeded alone, a matter of considerable pride for the Prof.

Lindemann asserted that America's McMahon Act, so resented by the British, had in any case been a failure, and said as much in a second note sent to Churchill on the same day. Trying to keep the Soviets locked out of the technology had simply not worked.

> Whilst it was thought that they only knew a little of what the English speaking people knew, it was reasonable to make considerable sacrifices to make sure that nothing whatever leaked through to them. But now that it is plain that the Russians know practically as much as we Anglo-Americans do, it is surely quite wrong to forego the advantage of collaboration on these vital intelligence questions in the hope of preventing a leakage of information to them.[21]

Churchill handwrote at the bottom a simple note: 'I agree'.[22]

One of the most important factors motivating Britain's behaviour was its relationship with the United States, which far outweighed its relationship with Australia in importance. The 'special relationship' with America was first named as such by Winston Churchill during his 'iron curtain' speech in 1946. This relationship, characterised by close military and diplomatic ties, was bruised by the 1946 McMahon Act that excluded Britain from US nuclear development. The shock of the sudden introduction of the Act at the end of the war was described by one scholar as 'a source of deep and abiding bitterness in Whitehall'.[23] The long process of bringing the two nations back together in this field proceeded with difficulty, but proceed it did. Britain needed America, and liked to think that the dependence was mutual. As writer Tim Sherratt put it, 'atomic energy policy within the British Commonwealth was always dependent upon the more crucial relationship with the United States that Britain so assiduously sought to maintain and extend'.[24] Since the US had a dim view of Australia's national security, so did the UK. Despite both countries needing Australian uranium, and Britain needing Australia's wide-open spaces for atomic testing, there was always a barrier between the UK and Australia, and many of the features of the era of British atomic testing in Australia were related to this. Australia was not in any sense an equal partner in the British nuclear tests, even though Britain constantly talked up the Commonwealth aspect. All pretence of a Commonwealth joint venture would be dropped quickly once America removed the legal barriers to nuclear technology co-operation in the late

1950s, and Britain abandoned its erstwhile Commonwealth collaborator with scarcely a backward glance. Even while Britain was cosying up to Australia, a distance was always maintained. At Operation Hurricane, Australia had minimal involvement and there was no formal oversight at all, only three notionally Australian observers, Alan Butement, Leslie Martin and Ernest Titterton, who wielded little actual power. That did change later, just before Operation Mosaic, with the advent of the Atomic Weapons Tests Safety Committee, but that organisation ultimately did little to protect Australia from the consequences of the tests.

A prime minister had won office in 1949 who was especially well disposed to any plea from Britain for assistance in its grand nuclear plans. The head of the recently created Liberal Party, Robert Menzies, would become a conservative legend, holding power for a record term (16 years) and embedding his sturdy, Anglophilic persona as a fixture in Australian public life. The warm feelings that Menzies had for Britain were not necessarily reciprocated. Neither Lindemann nor Churchill, nor any other British insider, particularly wanted anything to do with Australia. A tone of disdain or impatience with a country they considered lesser to their own comes through from time to time in the documents. None of the physical consequences of the decision to use Australian land were theirs to concern themselves with either, a lofty detachment that makes the British atomic tests in Australia even more tragic, given the long-term harms they caused.

Lindemann was all for the Americans repealing the McMahon Act so that bilateral nuclear research and development could pick up where it left off in 1945. However, even if the US did welcome Britain back into its nuclear tent he doubted that Britain could trust them to any extent. He simply did not believe that repealing the Act would solve the problem of the Americans wanting to be the dominant partner. 'Unless we can get complete [American] cooperation – and perhaps even if we can – I think we should try to build up a great Commonwealth enterprise to develop the civil and military uses of uranium.'[25] Nevertheless, despite elevated rhetoric about the Commonwealth frequently used by the British government, an attempt by Menzies to suggest an equal partnership between Britain and Australia on industrial atomic development was ignored. No such partnership was ever formed.[26] Evoking the Commonwealth to secure Australian co-operation may indeed have been a deliberate strategy by the British, employing 'complex imaginative fictions' around the threat to both Britain and its former colonies intended to consolidate the Commonwealth as a new Queen took the throne.[27] At that moment, with the death of King George VI in 1952, the same year as Hurricane, the fledgling Commonwealth stood on the brink, its future uncertain. Britain 'made strenuous efforts during the 1950s to consolidate the Commonwealth of Nations under the British nuclear umbrella'.[28] As history shows, Queen Elizabeth II championed and strengthened the Commonwealth throughout her long reign, which began just as her fading imperial nation was asserting its nuclear capabilities. The rhetoric of Commonwealth found a sympathetic ear in Robert Menzies.

Regardless of its reservations about the trustworthiness of the Americans in nuclear development, Britain did try extremely hard to overcome the barriers posed by the McMahon Act to work with them. For example, in July 1950, when plans to test in Australia were being evaluated, the British Joint Services Mission (BJSM), an organisation representing the British military services that had been operating in Washington under various names since 1941, tried to get the American Joint Chiefs of Staff (JCS) to agree to hosting British tests at the American site in the Marshall Islands, then known as Eniwetok (now Enewetak), in the Pacific. Lord Tedder, UK chief of the Air Staff, was in Washington at the BJSM and noted that the US was not averse to the idea:

> Reference use of Eniwetok range. I am informed that the J.C.S. have discussed this and would be prepared to entertain official request if made to them. Since I understand that the U.S. are now considering practicability of a land range they recommend that in asking for use of range facilities we do not specify Eniwetok. I am therefore putting in formal request for use of the range accordingly.[29]

When nothing further happened, an official at the Ministry of Defence in London wrote to Lord Tedder, saying that securing an agreement to test at an American site was 'needed to enable us to make necessary preparations, and unless it is forthcoming shortly we may be driven into costly examination of alternatives which may ultimately prove unnecessary'.[30] Finally in October,

US General Omar Bradley, chair of the JCS, replied to the BJSM request, and the news was not good:

> As you know the U.S. policy with reference to our relationships with the United Kingdom and Canada in the atomic energy field is currently under study. Pending the completion of this study I believe that it would be premature from the point of view of either the United States or the United Kingdom to authorise British use of the Eniwetok proving ground ... Meanwhile the U.S. is not in a position to consider the loan of the facilities to the United Kingdom.[31]

As the Americans backed further away from any association with the British test program under the terms of the McMahon Act and in the wake of the Klaus Fuchs spy scandal, all US test sites were off the table for the foreseeable future. The costly alternative of the Monte Bello Islands narrowed down to become the only option. On 31 August, a decision had been made (since there had been no word from the Americans) by the Ministerial Committee on Atomic Energy that Prime Minister Attlee should approach Robert Menzies to begin surveying the Monte Bello Islands, the first step towards establishing the viability of the site for atomic weapons testing. A briefing paper for Attlee confirmed that the initial approach to the Australian government would be at prime ministerial level and it was to be top secret. 'The matter is one of great secrecy, and I think that in your message to Mr. Menzies you will wish to emphasise this. I propose to instruct our High Commissioner in Canberra to take special security measures to safeguard all communications on this subject.'[32] A message from Attlee was prepared on 14 September and handed to Menzies in Canberra on 19 September. A telegram from the Commonwealth Relations Office (CRO) recorded the Australian's assent: 'Mr Menzies at once said that any special facilities which Aust might possess for this or similar purposes would of course be made available'.[33] The matter was not to be shared at this stage, as the telegram noted: '[Menzies] called Acting Minister for Defence (Mr McBride) into consultation and said that at this stage no one else would be informed except [Secretary of the Department of Defence Frederick] Shedden. Later if proposal eventuated he thought it would be desirable to take Leader of

Opposition into confidence'.[34] The next day Attlee replied with a simple message: 'I am most grateful to you for your help in this matter. We will be getting in touch with you about detailed arrangements as soon as possible'.[35] The tone of utmost secrecy was set, with Menzies readily agreeing to exclude most of his own Cabinet in the interests of keeping it all under wraps. As Australian author Robert Drewe remarked many years later, the 'Montebello bombs were an historic monument to the dominant political mentality of the day. And like many occurrences in 1950s and 60s Australia, their detonation was a case of *Keep quiet. Keep your head down. Don't rock the boat. The authorities must know what they're doing*'.[36]

Clement Attlee had good reason to believe that the Australians would not present any serious obstacles, and he was right. Now, after being humiliatingly snubbed yet again by the US, Attlee declared that it was 'clearly advisable if only as a precaution to consider possible alternative sites in British Commonwealth territory and to carry out a reconnaissance in the selected area'.[37] Menzies handwrote at the bottom of Attlee's message that he agreed 'in principle and to proposed detailed reconnaissance'.[38] Because the survey would require a large amount of input from both military and public service personnel, a 'cover story' was devised. Only a minuscule number of Australian officials knew the real reason for the activity at Monte Bello. The fake story was that the area was being surveyed to determine whether the islands were suitable as an instrumentation site as part of the long-range missiles testing site at Woomera in South Australia.[39] The cover plan was explicitly formulated 'to deceive subordinate officials making administrative arrangements for survey'.[40] The choice of cover story brought some complications, since those Australian officials actually involved with the Woomera missile project would obviously be perplexed by the story, unless they were inducted into the secret. Both the secretary of the Department of Supply, Jack Stevens, and the Department's chief scientist, Alan Butement, refused to co-operate unless they were told what was going on, especially as the ASIO security officer Leo Carter was going to be co-opted to the Hurricane project. Carter was an ASIO counter-espionage operative and former New South Wales police officer, who had a prominent role in the Petrov defection of the early 1950s. While ASIO was under the Prime Minister's Department, there was considerable liaison between the organisations. They were therefore brought on-board on a 'need-to-know' basis.[41] However, their minister, Howard Beale, was not. 'It was made plain that they were not at liberty to tell their Minister who would be acquainted with the project by the Prime Minister when he saw fit.'[42] When Beale eventually did find out, he was not best pleased, taking his exclusion as an insult.[43] For many months after the first approach from Attlee, only three people in Cabinet knew – Menzies, the defence minister, McBride, and Deputy Prime Minister and Treasurer Arthur Fadden.

Years after the Monte Bello tests, Operation Mosaic air task group commander Stewart 'Paddy' Menaul became a commentator on Britain's nuclear policy. Writing in an American strategic review journal in 1981, he alleged that the 'excessive' secrecy surrounding the decision for Britain to develop its own A-bomb was because Attlee didn't trust his own colleagues. Menaul asserted that 'some of [Attlee's] Cabinet colleagues and Labour members of Parliament ... had shown a marked preference for the type of socialism being practiced in the Soviet Union'.[44] Menaul's own political proclivities were very much at the other end of that Cold War spectrum, as a vociferous proponent of nuclear armament. After his role at Operation Mosaic, he had gone on to undertake the same role during Operation Buffalo at Maralinga in September and October 1956. He had been aboard the Valiant aircraft that conducted the only air-drop test during the entire British series in Australia, Buffalo 3.[45] He was among those officially involved with the tests who went to his grave convinced that they were justified and properly run.

As the Australian government fell in with British plans, there were many practical matters that needed to be set in motion, all under heavy secrecy. In October 1950, a photographic survey of the islands was undertaken by the RAAF.[46] The formal agreement for the Monte Bello Islands to be used for the first British test was struck between the UK and Australia in May 1951,[47] and that same month the British Hurricane Executive committee (known as HUREX) was established under the chairmanship of the deputy chief of Naval Staff, Vice Admiral Edward Evans-Lombe, 'to plan Hurricane as a joint operation under Naval command'.[48] A four-phase plan was agreed to help ensure that the tight timelines could be managed and all the logistical work completed. The preliminary phase would last until April 1952, during which Australia would complete surveys, lay navigational and berthing buoys and construct the first roads and ramps. In Phase I, from April to August 1952, two tank landing ships carrying engineers and equipment would prepare the site for instrumentation. Phase II, from August until the date of the test, was when the entire Hurricane force would be present at the islands; and Phase III, from the test date until January 1953, was when records and equipment would be collected.[49] The British test authorities now had a blueprint for the first test, and it was clear that there was an impossible amount of work to do.

The planning involved was complicated, not least because the British were doing most of the work from 15 000 kilometres away, and their test site was remote and uninhabited. Secrecy requirements and the covert, need-to-know ethos of the bomb authorities meant that few people knew exactly what was required and what was happening. The Monte Bello Islands had no infrastructure, and were known to have difficult weather conditions, being squarely in the torrid zone and subject to destructive cyclones. For Penney's part, this was a major test of his organisational and creative abilities – and his capacity to meet tight deadlines. He was not used to the maritime environment, although he had spent some time with the Americans in the Pacific before the McMahon Act forced him to leave in 1946. Nevertheless, he was up for the task.

Given the British imperative to better understand the risks associated with an atomic bomb detonated in a port, Operation Hurricane was conceived as a 'base surge' test. This kind of experiment was designed to investigate the tremendous upward movement of water and fission products from a water-based detonation. Related to this, it would also test what would happen to a small naval vessel when a bomb inside it was detonated. The looming and terrifying prospect of a port attack made this a priority. The site would not be a port, but the beautiful archipelago would have to serve as its proxy. In no way were the Monte Bello Islands like London, the city the British authorities most feared would be attacked. Nevertheless, they would provide the first opportunity for the British bomb-making authorities to fully prepare for their capital city to be targeted by the newest weapon of mass destruction. British eyes now turned with some trepidation to the pristine waters and craggy limestone coves of faraway islands.

CHAPTER 2

FIRST STIRRINGS OF A HURRICANE

> Hurricane ... was the skeleton key that would unlock all those American doors slammed shut when [atomic spy Klaus] Fuchs was exposed. Provided, of course, that the bomb went off.[1]
>
> HISTORIAN KEVIN RUANE

Britain had taken full advantage of having a large (former) empire, and searched extensively to find the right place to prove its atomic credentials. Robert Menzies, in his pride that Australia had been chosen, may not have been aware just how far and wide the British authorities had been looking for a location, by no means wedded to the idea of an Australian test: from the Bahamas to the Falklands to Somaliland.[2] According to the celebrated British defence journalist Chapman Pincher, America's rejection of British requests to test at existing US sites cost Britain £2 million and 18 months of effort, as it was forced to set up an entirely new site in Australia.[3]

The US was in no mood to openly work on atomic weapons with the British, given the damage done by the British spies in the Manhattan Project. However, the Americans did not unequivocally shut the door, either, giving Britain in turns hope and despair that it would be able to ever make use of the well-established American sites and American know-how. Despite multiple behind-the-scenes attempts to get the Americans to provide test sites, the situation increasingly appeared hopeless, with no firm answer from the other side of the Atlantic. By the time Attlee approached Menzies to access Australian territory, time was fast getting away and the quest for a British atomic weapon seemed to be losing ground.

The British planners had little fine detail to go on when considering the site chosen, a place about as far away from the origins of the weapon being tested as it was possible to venture. The British Admiralty had charts of the islands that were more than 100 years old, created when HMS *Beagle* visited in 1840. More up-to-date charts were needed, so Operation Epicure was established to survey the area. 'Epicure' was also the code name used in early secret communication between Britain and Australia on the test plans more broadly. The survey operation, undertaken in secrecy aboard the Australian ship HMAS *Karangi* and conducted by a team who arrived from the UK on 1 November 1950,[4] established the viability of Monte Bello for the tests and also collected a trove of biological samples that were sent to British museums for analysis. The Epicure report was submitted to the British government in January 1951, concluding that the Monte Bello Islands were suitable, but only at limited times during the year. The climatic conditions at these tropical islands meant that a test could only be conducted during October.[5]

William Penney reported to the UK government on the Epicure survey on 12 January 1951. He said that the Monte Bello site appeared suitable from the scientific and administrative perspectives, subject to further investigation of the depth of the water to accommodate the ships that would be present.[6] He added, somewhat gloomily, 'we know of no other site in which this trial could take place on the planned date'.[7] Penney outlined some of the provisions he saw as necessary for the operation, given the 'bare and dispersed' nature of the archipelago site:

(a) Recreational facilities, mostly ship borne, in order to maintain morale.
(b) Sufficient boats to ensure that the maximum effort can be deployed from the minimum scientific strength.
(c) Adequate communications to enable co-ordination of effort to be established and maintained.
(d) Fresh water, in which respect the expedition will have to be self supporting.[8]

While foreshadowing the infrastructure necessary for the operation, Penney was effectively flying blind. Nothing like Hurricane had been attempted before, so he was trying to invent a viable test program from the ground up. He determined that about 200 scientific and technical personnel from the UK would be needed on site, in addition to military personnel specially trained in radiation safety.[9] (Penney did not get everything he wanted – in the end only 100 scientists and technologists took part.)[10] He suggested that the complex and still not completely understood requirements might best be co-ordinated by the Ministry of Defence Committee on the Strategic Aspects of Atomic Energy under the chairmanship of the distinguished English chemist Sir Henry Tizard (a sometime critic of the British bomb program), with input needed potentially from the War Office, Home Office, Ministry of Works, Ministry of Health and elsewhere.[11] As plans developed, so did the administrative complexity, to the point of becoming almost impossibly cumbersome. The British civil service moved slowly at the best of times, and was a serious impediment now to fast completion of the Hurricane mission. One thing was needed immediately, though – a firm commitment to Monte Bello. Penney commented that the multiplying priorities, 'in addition to the main one of providing and fitting out ships for the operation, point to the need for a very early decision on whether the [trial] at MONTE BELLO can be approved, at least in principle, so that detailed planning of the operation may commence'.[12] The clock was ticking, and there was so much to do. Only in December 1951, though, after the October general election in Britain, was Monte Bello confirmed as the site for the test.[13]

Attlee communicated the outcome of Epicure to Robert Menzies in March of 1951, with a view to a test being held in October 1952. Because of the Australian elections of 11 May 1951, which returned the Menzies government for its second term, there was a slight delay in providing Australian approval. However, in light of the Epicure findings, Menzies authorised a detailed hydrographic survey of the islands, which was undertaken by HMAS *Warrego* in July and August 1951.[14]

The Hurricane Executive, HUREX, operating in London under Vice Admiral Edward Evans-Lombe, wanted to get the most out of the *Warrego* expedition, so requested that two British representatives join the ship.[15] The then-scientific officer of High Explosives Research, later the Atomic Weapons Research Establishment (AWRE), Noah Pearce, was assigned, along with Lieutenant Colonel AP Smith to get down to the detail of what was required. Pearce, who was also involved with a failed British clean-up later at Maralinga and authored a famously inaccurate report about the radiological state of the Maralinga range, was asked to work with Frank Hale, coordinating aerial radiological surveys of the Monte Bello area after the blast.[16] Pearce and Smith had several goals during their two-week 1951 visit, including the final selection of sites for structures and piers, reconnaissance of quarries, a tachymetric survey of the main base site on Hermite Island and collection of stone and sand samples.[17] In particular, Smith was tasked with gathering sand and rock to see if it might be possible to create on site the concrete needed for the rapid development of infrastructure on the islands. After Smith and Pearce's visit, 11 bags of samples were left with MacRobertson Miller Aviation in Perth, and from there passed to RAAF Base Pearce near Perth for testing.[18] At the RAAF base, seawater was mixed with the sand and rock to create concrete. Given the logistical difficulties of the islands, any avenues for sourcing materials and generally being self-sufficient on site were pursued.

Soon after the *Warrego* survey, the British government changed hands. Clement Attlee was out after six years, narrowly defeated by the Conservatives. Upon his return to the prime ministership in October 1951, Winston Churchill took great interest in the development of the British atomic weapon, constantly engaging with Frederick Lindemann on the subject and inserting himself – often unhelpfully – into planning. He initiated the short-lived APEX committee with the intention of having a mechanism for his own hands-on involvement, and exhibited far more enthusiasm for this committee than other members (especially Penney) who were press-ganged into attending. The opening two sentences of the minutes of the first meeting explicitly state why Hurricane was of such interest to the prime minister:

> The PRIME MINISTER said that he wished to be kept informed of the progress made in the preparations for the atomic test. The success of this test was of the utmost importance, since it would put the United Kingdom in a far better position to secure the full co-operation with the United States which was essential to the success of a strategic air plan in war.[19]

The first UK atomic weapons was something of an initiation rite that, all going well, would allow Britain to become one of the select few inside the nuclear club. Churchill needed Hurricane to be successful, not just because of the huge expense that had gone into its development, but also to use it as a way to forge a new atomic-era relationship with the United States.

Once Monte Bello was deemed suitable, a huge military operation to build the infrastructure on the islands swung into action, and a cat-and-mouse game with local media began in earnest. The British wanted to hold off on any public discussion of their first test until close to the time. The official public announcement was not made until May 1952, but speculation proliferated up and down the Western Australian coast, as the volume of naval traffic increased.[20] Once the news was official, the media noted – with some justification – that Winston Churchill had put himself at the centre of the project. A typical news report, with a dateline of London, led with: 'The Prime Minister (Mr. Churchill) had assumed direct and complete control over all arrangements for the detonation of Britain's first atomic weapons off the north-west of Australia this year – probably in September – London newspapers said today'.[21] The story continued that the project was so secret 'that even high Admiralty officials were being kept in the dark, and day-to-day problems requiring solutions were being submitted direct to the Prime Minister'.[22] While it is doubtful that Churchill was dealing with all the many practical problems in this way, the vanity project of the APEX committee was driven by the British prime minister's deep interest in the atomic bomb. Churchill did indeed make a ruling that nothing would be said in public about the operation that had not been directly cleared by him. Minutes of the third APEX meeting spell this out: Churchill declared that he would 'personally approve any communiques or statements which it was desired to issue to the Press. Apart from these, nothing whatever should be said to the Press and nobody should be given any discretion at all to make any statement of any kind'.[23] The hatches were to be battened down for this first British A-bomb. Years later, William Penney would remark, in an obituary for his offsider William Cook, that Churchill and Frederick Lindemann 'were personally masterminding national Defence policy about nuclear weapons'.[24]

That first APEX meeting was small and prestigious, featuring the most important figures in the political and scientific development of the British bomb. Aside from Churchill there was Minister of Defence Earl Alexander of Tunis, Paymaster-General Frederick Lindemann (Lord Cherwell), Minister of Supply Duncan Sandys, Sir Roger Makins of the Foreign Office, Neil Pritchard from the Commonwealth Relations Office (later UK deputy high commissioner to Australia), Vice Admiral Edward Evans-Lombe of the Admiralty (soon to be named head of the Hurricane Executive, HUREX), Lieutenant General Sir Frederick Morgan from the Ministry of Supply (who had succeeded Lord Portal as Controller of Atomic Energy), Rear Admiral Arthur Torlesse of the Admiralty, who would command the naval forces at Monte Bello, and Dr William Penney of the Ministry of Supply, inventor of the British A-bomb.[25]

Lindemann did the heavy lifting to create APEX, five months after the election, at Churchill's insistence.[26] Churchill chose the name APEX because 'The other day I saw a cartoon – two ape-like creatures in a blasted landscape, the male approaching the female with an ingratiating look on his face, and she saying, "No! We are not going to start all that over again!"'. That explained the APE. 'The X, of course, stands for the unknown quantity.'[27] Most of the high-level group who attended were quietly getting on with their own planning activities separately, without the presence of Churchill.

APEX met just four times, as its functions quickly were being duplicated by HUREX. Nevertheless, the meetings are interesting windows into the thinking at the top of the British government as it prepared for a momentous change to its defence strategy. Despite the high-level discussion, APEX was largely a waste of time, according to author Kevin Ruane: 'from the start APEX was resented as an unwelcome distraction by those forced to attend to satisfy the curiosity of a Prime Minister uninterested in reading briefing papers'.[28]

The second meeting was chaired by Churchill on 15 April 1952 and attended by Sir John Cockcroft, head of the Atomic Energy Research Establishment (AERE) at Harwell. Cockcroft told the meeting that, as a result of his conversations in the United States,

> he hoped that co-operation in atomic energy matters would now become easier. He expected that during the remainder of 1952 there would be a marked improvement in Anglo-American co-operation in intelligence matters and some small improvement in technical

> co-operation. Co-operation would doubtless improve in 1953 if our test were successful.[29]

He assured the meeting that the contretemps over the Fuchs espionage matter was being smoothed over and the Americans were watching UK weapons development with interest. The minutes mention the 'friendlier atmosphere'[30] since the Fuchs-inspired fear and suspicion had been somewhat allayed.

Meanwhile HUREX, made up of senior service and departmental officials, was orchestrating plans apace. The committee was required to co-ordinate action under the direction of the UK Ministry of Supply, the operation's controlling department.[31] Penney appointed ballistics specialist Dr Leonard Tyte as the technical director for Hurricane. HUREX liaised directly with the Australian Hurricane Panel, chaired by Captain Alan McNicoll. Also on the Australian panel was Colonel Charles Spry, inaugural chief of the new Australian Security Intelligence Organisation, ASIO.

Operation Hurricane was subject to the newly established Australian D-notice ('Defence notice') system in which media were willing participants in restrictions on reporting. This self-censorship of media reporting meant that the true nature of the tests was not revealed to the Australian public. Media outlets at the time seemed to take the view that secrecy was justified. As an editorial in *The Age* stated, it was 'better that we should know too little than run the danger of putting information into the hands of those who, at this moment, are, no doubt, seeking to discover the scope and potentialities of the Monte Bello experiments'.[32] The Royal Commission into the British nuclear tests would later conclude that there was 'virtually complete government control of the Australian media reporting of the Hurricane test and the lead-up to it, thus ensuring that the Australian news media reported only what the UK government wished'.[33]

One of the earliest atomic tests media releases issued by the Australian government dates from two months before Hurricane. This bland, two-paragraph document announced the attachment to the project of Professor Ernest Titterton.[34] Titterton, who went on to have a highly controversial involvement with the entire British atomic test program in Australia, would be vilified during the Royal Commission during the 1980s for allowing his loyalties to the British nuclear elite to interfere with his ostensible role of providing impartial advice to the Australian government. The two other Australian observers at Hurricane, the Australian defence scientist Leslie Martin and the chief scientist at the Department of Supply, Alan Butement, were not announced publicly and their roles were shadowy and ambiguous.

Obsessive British secrecy dominated in the build-up to Operation Hurricane. The stakes were high, and the British government did not trust its own media, let alone that of Australia. The plan to detonate a British atomic weapon was announced simultaneously in Britain and Australia on 14/15 May 1952, some 20 months after Attlee had first approached Menzies. When the Brisbane *Courier-Mail* newspaper ran a story on 30 May 1952 about the operation's flagship, HMS *Campania*, leaving dock in the UK for its voyage to the Monte Bello Islands, concerns were raised high up in the British government. A telegram from the UK high commissioner in Canberra to the secretary of state for Commonwealth relations, the Marquess of Salisbury, indicated that Arthur Fadden was equally concerned, given that D-notices had been enacted. 'Acting Prime Minister is somewhat perturbed at release of this report from London since he had just requested General Managers of Australian newspapers to refrain from publishing any reports bearing on atomic weapons other than those released by Government.'[35] The story did not contain much information, and the telegram conceded that the only new fact concerned the date that *Campania* had sailed. Nevertheless, since the British government had been so strict with the press, it was furious that the Australian press appeared to have information not officially supplied in Britain. The UK high commissioner wrote that publication of this story 'has stimulated fears already voiced privately that Government here may be exposed to criticism for taking too strong a line with press'.[36]

Britain was horrified at the thought of the Soviet Union exercising its espionage prowess the way it did during the Manhattan Project, and drew lessons from its activities during the Second World War as it developed its secrecy plans. Operation Tigress was developed in the lead-up to Hurricane to deceive the Soviet Union by planting misinformation about the forthcoming atomic test, largely to do with the timing of the operation, technical complications and movement of key scientists.[37] Indeed, Penney alluded to this in a note he sent to Leslie Martin in July 1952, explaining why he was travelling to Monte Bello via Woomera. He said, '[y]ou may be puzzled by my going first to Woomera, but I am being used as a pawn in a rather high-powered Intelligence operation and I am not my own master'.[38] Tigress built upon similar exercises during the war, such as the London Controlling Section, which had masterminded large-scale deception operations.[39] Much of this misinformation was disseminated through oblivious media outlets.[40]

Hosting a highly secret weapons development operation at Australian islands was clearly going to require some legal considerations. To secure the Monte Bello Islands for Operation Hurricane, the Australian government initially planned to extend the *Supply and Development Act 1939–48*, prohibiting access to large areas, including the islands and Rowley Shoals to the north-east. This measure was opposed by Britain, which was sensitive about any country overstepping territorial limits. The CRO sent a top-secret telegram to the UK high commissioner in Canberra, expressing fears that there 'may be difficulties in international law if Australian government proclaim areas as wide as those which they propose'.[41] The message cited the American practice at its Pacific test location, Eniwetok, which involved initially closing only the atoll and its territorial waters, and widening the danger zone for the test itself. The British specifically and emphatically wanted to ensure that journalists had no access to the islands, and this could be achieved

> by proclaiming at this stage merely Monte Bello Islands and Rowley Shoals and their territorial waters (the extent of which should be left vague, thereby avoiding the controversial question of their exact limits) and we urge that Australians should amend notification accordingly. We shall propose establishing a danger area (on lines of Eniwetok procedure) nearer the date of the trial.[42]

Despite British concerns about prohibiting large areas outside territorial waters, Australia actually went further and devised new legislation. The *Defence (Special Undertakings) Act 1952* placed strict prohibitions on who could enter the large proscribed areas. The new bill proposed prohibiting an area of over 144 kilometres diameter, centred on Flag Island in the

Monte Bello group. A minute paper for the defence minister, Philip McBride, on the draft bill in May 1952 said that while this went well beyond normal territorial waters 3 miles from land, 'it is understood that the position may be covered by letters patent issued last century by which the boundaries of Western Australia were fixed in the ocean at a point well beyond the Monte Bello Islands'.[43] McBride said that the questions of international law could be further examined by the Attorney-General's Department. Cabinet approved the bill later in May. It contained explicit reference to the Monte Bello Islands, but was 'capable of application to other operations'.[44] The Act gave the defence minister sweeping powers to declare any area in Australia prohibited, and the right to detain, search and remove any person found in the declared areas. The bill was passed into law on 10 June 1952. Only people with a permit could enter, and the only officials who could issue permits were the minister for defence, the chief of Naval Staff, the deputy chief of Naval Staff and the naval officer-in-charge West Australian Area.[45] These new powers were unprecedented in Australia, and have yet to be repealed.

The UK High Commission was advised of the details of the new Act by Vice Admiral Alan McNicoll, deputy chief of Naval Staff and head of the Australian Hurricane Panel. Britain was concerned that the Act broke international law in encompassing areas that were normally considered outside Australia's territorial control, so the advice was intended to allay concerns. McNicoll said that although 'the area does contain open water, the whole is within the limits of Western Australia, as defined in Letters Patent of 1900 which have not as yet been challenged'.[46] The note spelled out some of the implications of the new Act, specifically that it would restrict flying over a huge area of Western Australia, from Carnarvon to Broome, 'making it extremely difficult for any charter aircraft to take off, visit the islands and return. This will not in any way hamper the normal flying in the area'.[47]

Consternation grew in London. The British deputy high commissioner in Australia, Ben Cockram, wrote to the assistant secretary in the Prime Minister's Department, Allan McKnight, indicating that the British now regretted not pressing their case more forcefully before the Act was passed. Cockram said, 'I have now received a further telegram from London expressing regret that it had not been appreciated at the time that the provisions of the Bill would have the effect of permitting areas to be proclaimed which might extend beyond the limits of territorial waters'.[48] Clearly the British government, which had had reservations about earlier plans under the old *Supply and Development Act*, was even more worried now.

> The United Kingdom authorities are anxious that their previous absence of comment should not be considered as committing them in any way to the view that it is in international law possible for a Government to exercise sovereignty outside the limits of its territorial waters, and they have enquired as to the exact provisions of the law in this respect.[49]

Cockram followed up with another letter to McKnight on 1 July 1952, saying that the British government now feared that both governments could run into 'serious difficulties in international law'.[50] The language was getting stronger, with Cockram saying that trying to assert sovereignty outside of territorial waters could 'seriously embarrass' the UK, given the stand it had taken on the 3-mile limit in other parts of the world, notably Iceland.[51] He suggested following the lead of the Americans at Eniwetok and just declare hazardous areas off-limits closer to the time of the test, and leave it at that.

The concerns were escalated to the secretary of the Prime Minister's Department and Cabinet secretary, Allen Brown. He replied to Cockram by subtly indicating that Australia was a sovereign nation with all the legislative powers enjoyed by its former coloniser, and therefore had every right to enact such legislation.[52] He did not commit to making any variation to the legislation based on British concerns, but merely to look into the questions Cockram had raised. The whole mess was handed over to the solicitor-general, Sir Kenneth Bailey, who made a formal response on 18 July 1952. Bailey concurred with Britain in part, saying that he understood their 'embarrassment' and shared their concerns about the press getting access to the site. 'I have myself been much disturbed by the publication of press messages, purporting to come from London, which contain almost an incitement to anyone who could hire a foreign vessel to try out the validity of the security measures provided for in the Act.'[53] He did indicate that it would be difficult to be sure everyone who needed to be excluded was actually excluded without such strong legislation. In the end, however, he deemed this a policy issue rather than a legal one, and suggested that it be referred to the Hurricane Panel. The involvement of the solicitor-general in advising the government and the navy was picked up by the media. A story in the *Melbourne Herald* on 2 August 1952 flagged that defence authorities were concerned that a foreign country 'may try to pierce the security screen around the Monte Bello Islands during the atom tests'.[54] The article suggested that the new measures were particularly meant to guard against submarines (not journalists).

The Hurricane Panel duly considered the implications of the new Act and the British concerns, and came down on the side of the legislation as written. Given that the Australian government felt that it had some responsibility for the security of Hurricane on its territory, 'the Panel feels that it is difficult to see how that responsibility can be discharged if the maximum legal authority is not to be exercised'.[55] The legislation stood, unamended.

While the two governments were attempting to find the best way to maintain the security of the islands, William Penney and his team were preoccupied with more technical concerns. Radiation safety was among them, despite the fact that the British nuclear tests in Australia would be characterised by their lax approach. In these matters, the British were intent on finding the most acceptable guidelines that did not impede the rush to get the weapons tested. In the early 1950s, only a few years after the development of atomic weapons, safety standards for those working in weapons production and testing were still being established. Britain did not have its own regulations at the time, although it was able to draw upon the International Commission on Radiation Protection (ICRP), which had set safety standards for ionising radiation since 1928. Ionising radiation, released by atomic and hydrogen weapons, is capable of changing the cells of a living body by knocking electrons out of stable atoms, thereby damaging DNA. The three main kinds of ionising radiation are alpha, beta and gamma. The initial blast from a nuclear weapon is dominated by gamma radiation, while alpha and most beta emissions become a hazard if radioactive material is inhaled or ingested since they do not directly penetrate the body.[56] The relevant ICRP standards published in 1950 related to workers regularly exposed to uniform, low levels of radiation through their daily work. These standards were unsuitable for those present at an atomic test, when there would be a huge initial outpouring of radiation that tailed off quickly. There would be no 'uniform' dose at this first British test.[57]

Naval Task Force Commander Arthur Torlesse formally asked Penney to establish safety parameters for the trial. In his letter to Penney, Torlesse said that while safety was of the highest priority, 'some degree of risk must be run by some people if we are to achieve the full purpose of the trial'.[58] The nuance in this form of words was significant. Penney subsequently sought advice from the British Medical Research Council's Panel on Atomic Biological Effects, headed by Sir Harold Himsworth. This panel had been working on the problem of maximum allowable doses of gamma radiation, an especially dangerous form because its penetrative powers mean that it can affect bodily organs from outside the body. Beta is equally hazardous, but only when inhaled or ingested. Based on this advice, Penney and his colleagues settled on permissible dose rates for both beta and gamma radiation.[59] Penney's initial safety proposals reflected these. However, when these guidelines started to look too restrictive, and it became clear that perhaps volunteers could be asked to put themselves at greater risk if it would benefit the test, Penney made a change. His subsequent proposal, submitted to the Ministry of Supply and the Admiralty, asked for a higher rate of allowable beta radiation (the gamma dose was unchanged). He sought approval to redefine the acceptable dosages, based on varying ratios of beta and gamma radiation, and because the light clothing to be worn in the tropical environment would not provide much protection.[60] Penney created three categories for exposure: normal working rate, lower integrated dose and higher integrated dose. The latter two gave Penney and Torlesse more leeway in directing personnel to enter contaminated areas after the test. The lower integrated dose would 'be applied only with the express permission of the Radiation Safety Officer, which will be given only where he regards it as necessary for the smooth running of the operation',[61] and anyone exposed at this level would not be able to be exposed for the rest of the operation. The higher integrated dose would 'be applied only in cases of extreme urgency in order to recover vital records which might otherwise be lost', and anyone receiving this dose would have to stay away from any possibility of further exposure for 12 months.[62] These new standards were accepted by the British government and written into the Hurricane trial orders.[63] With these standards in place, it became possible to put test personnel into harm's way, with full official consent. The cost of the operation and its high political stakes made for powerful justifications to begin the program of British nuclear testing with a safety regime that had room for interpretation.

CHAPTER 3
NAVAL OPERATIONS

> Several hundred people living in the North-West had an idea that the Navy was undertaking some kind of secret work when they saw H.M.A.S. *Mildura* and H.M.A.S. *Karangi* operating around Onslow. However, they maintained their silence and the R.A.N. was able to keep its most strictly guarded secret.[1]
>
> *WEST AUSTRALIAN*, APRIL 1952

> All ships had left the lagoon the previous week, giving us a very doomed feeling as even the little Aussie lighters chugged past to seaward, and tension began to heighten as more boffins descended to make final adjustments and cold-bloodedly asked which bits of our ship they could take away as souvenirs.[2]
>
> *THE PLYM PUDDING*, A NEWSLETTER OF THE LAST DAYS OF HMS *PLYM*

As the bureaucratic and scientific teams worked hurriedly to prepare for Operation Hurricane, the naval side also geared up. This was to be a major undertaking, the largest British naval operation since the war. Hurricane's flagship, the converted aircraft carrier HMS *Campania*, made the trip to the Monte Bello Islands in company with two other Royal Navy vessels, HMS *Tracker* and HMS *Plym*. *Campania* and *Tracker* would form the Special Squadron when united with HMS *Zeebrugge* and HMS *Narvik*, which had gone before and had transported Royal Engineers and their heavy equipment out to Monte Bello. *Zeebrugge* and *Narvik* were both 'landing ships, tank' (LSTs), and together they were designated the '4th Landing Ship Squadron' for Hurricane. The small River-class frigate

Plym was the sacrificial ship, doomed to be blown apart by the Blue Danube device detonated in its hull. The shorthand term used was 'TV', short for target vessel.[3] In keeping with British attempts at diverting attention, TV was deliberately chosen as cover, to encourage the belief that the ship had something to do with television. Official historian Margaret Gowing said that an unneeded gantry covered in antennae was added to the ship to assist with the deception.[4]

Rear Admiral Arthur Torlesse, Commander Task Force (CTF), hoisted his flag on *Campania* on 14 May 1952, coinciding with the first announcements of the test in the British and Australian media (15 May in Australia). Only then were the crew on board the flotilla told of their mission, in secrecy measures that had been decided at the first meeting of Churchill's APEX committee in March.[5] *Campania* and *Plym* had already held a secret rehearsal in the Thames estuary.[6] The long voyage, about eight weeks, took the ships around the Cape of Good Hope instead of transiting through the Suez Canal, which was deemed to be unsafe for this secret mission.

For its final voyage, HMS *Plym* was captained by Lieutenant Commander Peter Draisey. He was joined by 87 crew members, five officers and four scientists.[7] *Plym* was a relatively young ship – only ten years old – but was no longer needed after the war. She carried in her hold the main components of the Blue Danube device, without the plutonium core, although most of the ship's company were unaware that the device was on board. The core was flown out to the Monte Bellos by Sunderland seaplane two weeks before the scheduled blast, accompanied by AWRE scientist WJ Moyce.[8] The air trip took three days, and the core was packed into a cork box that had a bag of dye attached. If the plane crashed, the dye would be visible in the water.[9] *Plym* rendezvoused with *Campania*, and the health control ship HMS *Tracker*, to form a mini flotilla for the journey to Australia. *Tracker* parted company with the other two ships to pass through the Suez Canal.[10]

On their earlier voyage out to Monte Bello, tragedy had visited *Zeebrugge* and *Narvik*. When both ships berthed at the Cocos Islands in the Indian Ocean, then administered by the British as part of Singapore but in 1955 to become an Australian territory, the crew were glad of the glorious tropical environment in which to recuperate during an arduous trip. A detachment of Australia's RAAF No. 2 Airfield Construction Squadron (2ACS) was stationed there, building an airstrip as part of the effort to enable flights between Australia and South Africa.[11] This doughty outfit had earlier built the Woomera rocket range in South Australia. The 2ACS commanding officer was Percival 'Nobby' Lings, who gave permission to Captain Colville, commander of HMS *Zeebrugge* and senior officer of the 4th Landing Ship Squadron, to let his men come ashore for some socialising. They did so on Sunday 6 April 1952.[12] What happened next was horrific. Despite orders that the visiting troops were not allowed to swim, five British sailors jumped into the enticing waters. When they got into difficulty in the treacherous surf and undertow on the reef, members of 2ACS went to their aid with immense bravery. In the desperate and hours-long struggle to save the five British sailors, two Australians perished – Michael Rowan and Peter Eccleston.[13] One of the British sailors, John Atkinson, also died.[14] In September 1952, the Australian minister for civil aviation, Larry Anthony, came to the Cocos Islands to unveil a plaque erected in memory of the three servicemen who lost their lives.[15] The newly ascended monarch, Queen Elizabeth II, who would be crowned in 1953, recognised the heroic efforts of RAAF Corporal Bob Stewart with the award of the George Medal, and RAAF Corporal John Kelly was recognised with the British Empire Medal.[16] Stewart, a 24-year-old Gold Coast lifesaver, 'narrowly missed being pounded to death on the coral reef, and after bringing in three of the Navy men, was swept about two miles down the coast before he could regain the shore'.[17] Rowan and Eccleston were posthumously awarded the Queen's Commendation for Brave Conduct.[18] The whole sorry saga hung a pall over *Zeebrugge*'s and *Narvik*'s onward journey to Fremantle then Monte Bello, a precursor to the sorrow that the nuclear test program in Australia would bring to generations of Australians.

Hurricane was a military operation, but one driven by scientific and political imperatives. Its planning involved multiple agencies and individuals, to deal with the almost overwhelming logistics. One thing that had not been planned for were the clashes in culture as the military personnel led by Admiral Torlesse had to deal with the scientific and technical staff overseen by Leonard Tyte, under the direction of William Penney, a pairing neither side was used to. Unfortunately, Tyte travelled with Torlesse aboard HMS *Campania*, for the uncomfortable two-month journey. Torlesse and Tyte disliked each other immensely. The trip aboard

Campania became notorious for the lack of harmony between the scientists and the navy personnel. Despite the efforts of the naval liaison officer, Captain Pat Cooper, the two factions could not get on. Tyte and Torlesse had a monumental personality clash, and their enmity filtered down to both groups. Official historian Margaret Gowing said that the 'impact of two different worlds – the conservative, traditional and highly disciplined Navy and the informal, individualistic "boffins" – was sharp'.[19] Australian author Paul Grace put it more bluntly: 'Admiral Torlesse and Dr Tyte bickered constantly. The sailor thought the scientist was thin-skinned, ungrateful and overly critical, while the scientist thought the sailor was pompous, snobbish and inflexible'.[20] The petty differences played out in various ways. On one occasion, Gowing relates, all on board were directed to wear a tie to the Sunday night film screening. Most of the scientists turned up in ties but no shirts, a sight guaranteed to annoy the sailors.[21]

Never had the Royal Navy been required to transport so many civilians, and the stresses and strains were many. Captain Cooper wrote a frank account of the problems in 1954. Upon delivery of this report, there was some angst within AWRE about whether it could be seen to 'rake up old sores', as described by Penney's administrative deputy Admiral PW Brooking in a brief note attached to the report.[22] Brooking said that 'much friction could have been avoided if better relationship had been reached at the top, that is between R. A. Torlesse and Dr. Tyte'.[23] As Cooper said, 'the responsibilities given to the Naval Commander [Torlesse] ... were never more than grudgingly accepted by [Tyte] ... and, from start to finish, caused a sense of discord'.[24] Cooper rather dryly suggested air travel for the scientific staff for any future operation.[25]

Tyte wrote his own diary, and was harsh in his criticism and colourful in his language. Brooking also attached a note to the diary to the head of HUREX, Edward Evans-Lombe, describing the disharmony as 'a basic weakness' of Hurricane planning and execution.[26] He said, 'we clearly do not want this matter dragged into the limelight'.[27] Tyte's diary drips with contempt, to the point of describing Torlesse as 'a very stupid old man'.[28] He was angry at the lack of empathy of the navy personnel about the fact that scientists were not used to living and working on a naval vessel, writing 'there are some duty officers and boat staff who regard our amateur efforts as subjects for mirth rather than matters on which they can help'.[29] He also wrote that life 'would be much easier if the Naval Staff Officers we have here didn't interfere with things they don't understand and believe me that's quite a bagful'.[30] While Captain Cooper clearly had uncommon levels of patience in his efforts to promote goodwill, there was precious little to be had.

A member of the British scientific staff, JJ McEnhill, was the designated historian on the mission and provided a detailed narrative of the voyage of HMS *Campania* out to Monte Bello in 1952. Later, at the Australian Royal Commission held in the mid-1980s to uncover what had happened during the British atomic tests, this narrative was presented to show something of the human aspects of the operation. McEnhill said that 'by blending together the domestic and the technical it is hoped that the reader may obtain some overall impression of the operation, of the conditions prevailing and of their impact and effect on the scientific staff'.[31] He gave his own insights into the many frayed edges between military and non-military personnel on that long, fraught trip from one side of the world to the other. From mishandled scientific stores to the kind of heat British scientists and mariners had seldom experienced, to personality clashes and sobering lectures about the nature of atomic bombs, the journey was a memorable one.

Once at the islands, the scientific staff were keen to get off the ships and firmly place their feet on dry land, as far away as possible from the sailors. Various tented shore camps started springing up around the islands. McEnhill said the civilians longed to feel that they were 'their own masters, where they created their own mode of living and where they had no occasion to come into contact or require the direct assistance of the Naval Staff'.[32] McEnhill described, euphemistically compared to the Cooper and Tyte accounts, a 'strong undercurrent of strained relations' between two camps.[33] While the scientists appreciated the efforts of the Royal Engineers to prepare for the test, they did not feel the same way about naval personnel. 'A horde of khaki clad civilians, some less clumsy than others, who were daily poured into small boats certainly aroused Naval amusement, but often little more than that.'[34] One day *Campania*'s executive officer, Commander Douglas Bromley, had to jump into the sea to save one of those clumsy civilians, a member of the radiation hazard team, who had slipped off the gangway between the *Campania* and one of the small boats used to ferry personnel around the islands.[35] All of this was played out against a backdrop of extreme time pressure, with teams having even less time than they had budgeted for once they were on the islands, and unforeseen problems like sand blowing into delicate equipment, none of which improved anyone's mood. Also, the poor communication between the military services and the Ministry of Supply was an irritant, causing unnecessary problems.[36] Mediocre food and plagues of flies at the shore camps did not help morale either.

During Hurricane, the Monte Bello Islands became a military outpost, with strict rules around who could enter the area. Orders came from HUREX, via the Admiralty, on 11 September 1952, passed to Torlesse and onwards to commanders of the Australian ships, that any vessel breaching the prohibited area could be boarded. Furthermore, 'use of force if necessary in period immediately prior to trial will be supported here on grounds of safety of ship and her crew'.[37] This action was confined to the period from D-1, the day before detonation. 'If recalcitrant vessel is encountered any earlier time in danger area but outside territorial waters the incident is to be reported ... but force or armed repeated armed boarding party are not to be used without higher authority.'[38]

The preparation of the site had begun with the arrival of an advance party of No. 5 RAAF Airfield Construction Squadron in August 1951.[39] Bringing their heavy construction equipment with them by rail and road from their base at Williamtown, New South Wales, via Geraldton, they passed through Onslow on their way out to the islands. Most of the infrastructure work was on Hermite Island, which was to be the main shore site. Two tented camps were established once the British ships arrived carrying most of the scientists, one on Hermite and one on Trimouille.[40] The Australians got a head-start on the British Royal Engineers 71 Field Squadron, whose main group of 200 arrived in April 1952. By the time the scientists arrived on *Campania* on 8 August 1952, the set-up was impressive. According to Captain Cooper, the command centre on Hermite Island 'seemed much like the laboratories at home'.[41] As the scientists prepared the coming trial, the weather was largely fine and sunny with a light breeze, although the swell in the parting pool was 'capricious and varied with considerable rapidity'.[42]

The Royal Australian Navy played a pivotal role in Operation Hurricane, at some cost to Australia. Expenditure on RAN services such as food, fuel, wharfage, navigational aids and moorings was estimated at over £128 000,[43] roughly $5 million in today's money. The cost of personnel involved in the various surveys ahead of Hurricane, and the involvement of HMAS *Karangi*, HMAS *Koala*, HMAS *Mildura*, HMAS *Hawkesbury* and numerous other RAN ships was not charged to the test account.[44] Likewise, HMA ships *Sydney*, *Tobruk*, *Macquarie* and *Murchison* carried out air, surface and anti-submarine patrols around the Monte Bellos just before and after Hurricane, and these costs were absorbed by the Australian government.[45] HMAS *Culgoa* became the 'weather ship' for the period before the test, based about 500 kilometres from Monte Bello, out in the Indian Ocean.[46] HMAS *Warrego* and HMAS *Karangi*, although they had been involved in surveying the islands, were not present for Hurricane. Once the British ships were positioned at Monte Bello, joined by ships of the Royal Australian Navy on 27 September, they were provided with fuel by the oil tanker *Wave King*, which had journeyed from Sri Lanka to the site.[47]

The Australian Army had a smaller role to play, but still had a significant presence in support of major infrastructure development and security. The army set up a command base at Broome to provide accommodation facilities for the temporary RAAF base established in this small town in north-west Western Australia.[48] There was also an army transit camp at the town of Onslow, which at that time had a population of about 400. The army also provided a small security team on Monte Bello for several months after the Hurricane explosion.[49] The cost to Australia of all three fighting services that participated in Operation Hurricane was £201 662, a precise figure worked out later by the Department of Defence, which excluded pay and maintenance costs.[50]

As the navies of two countries converged on the islands, and their presence started to be noticed, another kind of drama started to unfold. The story of the media coverage of the tests is a remarkable tale of journalistic determination. Despite some lobbying, the media had been banned from joining the official party at the islands. As the bush telegraph along the Western Australian coast started to pick up on unusual maritime activity, Norm Milne, a waterfront reporter from the *West Australian*, was assigned to follow the story. And follow it he did, with a finely tuned nose for news. He had already seen *Narvik* and *Zeebrugge*, along with two small

Australian ships, arrive at the port of Fremantle, and noted the 'light security' surrounding their presence.[51] The British ships arrived in April, only a week after the drowning tragedy at Cocos Islands. Milne could see that *Narvik*, with its landing craft clearly visible, was set up for a major military operation, most likely the nuclear test, he surmised. While it was known that Britain would soon test its own atomic weapon somewhere in Australia, the news that it would be at the Monte Bellos would only be released to the press on 15 May. Milne's handwritten notes record that he got an inkling of the island destination for the British ships in classic newshound style – he talked to a barmaid at a pub in Fremantle. She told him that her boyfriend, a Royal Navy petty officer, was heading north to some islands. 'Got hold of charts – only 1898 available and took educated guess that it was Monte Bellos,' Milne wrote.[52] He put his theory to the naval officer-in-charge of the West Australian Area, Captain Bryce Morris, 'and all hell broke loose'.[53] Milne claimed that Robert Menzies himself wrote to his managing editor, Jim Macartney, asking him not to publish anything. He was also visited by 'a top security Colonel', who tried to convince him he was wrong.[54] Milne wrote in his notes, with apparent satisfaction, '[n]aturally it was splashed across the front page'.[55] Those handwritten notes also contained details he had gleaned about the islands to share with his readers. Two aerial photos of the islands appeared on the front page, with a brief caption saying that this was where 'it is believed, Anglo-Australian atomic tests may take place'.[56] On page ten, more information was provided, describing the physical characteristics of the Monte Bellos, based on the notes he had made at the time the saw the ships in Fremantle.[57]

This was a huge story, and journalistic blood was up. Soon after identifying the Monte Bellos as the likely venue for the tests, but before the *Defence (Special Undertakings) Act* was invoked to shut down the area to any non-authorised person, Milne joined fellow *West Australian* reporter and expedition leader Jack Nicoll, photographer Harold Rudinger and Jack Coulter from the *Daily News*, on a road and sea trip up the coast. Nicoll was an ex-navy officer, Coulter would live to 93 and win two Walkley awards for his police reporting, while Rudinger was a technical innovator who helped assemble the special camera equipment that would later capture dramatic vision of the British bomb.[58] The quartet travelled the 3200-kilometre round trip in a new (though soon to be wrecked) Ford from Perth to Onslow, a journey of considerable incident and the stuff of many a pub yarn later. *The Journalist* newspaper later recounted that two of the men shared a double bed in the bridal suite of a motel in a one-horse town along the way.[59] In the small mainland staging post of Onslow, the journos joined a crew of two on board the 22-foot yacht *Thelma*, and a legend of Australian journalism was born. As *The Journalist* article stated, 'They can rightfully claim to be the last civilians to see the Montebellos before the atomic tests do whatever they will do to them'.[60] They travelled out to the islands, in great discomfort but with their sense of adventure undiminished, catching crayfish and other seafood and poking around the complex geography of the archipelago. To their amazement and journalistic satisfaction, they discovered the ships of the Operation Hurricane Landing Squadron, which Milne had seen at Fremantle port, at anchor at the islands, with HMAS *Karangi*.[61] As soon as they were spotted, the journalists were directed to leave by Commonwealth security officers, but they declined, citing the undeniable fact that the official restrictions had not yet taken effect.[62] Coulter's story ran in the *Daily News* on 29 April, with a banner across the top saying that this 'story and earlier stories on the Australian atom-weapon test have been voluntarily referred to an appropriate official authority so that they comply with security precautions'.[63] Coulter speculated that a permanent observation base was being established at the islands, suggesting that it could be used for observing atomic tests on nearby Barrow Island. He reported that, other than 'saying that the Pressmen would be wasting their time by going alongside, the Navy placed no impediment in the way of the visitors'.[64] The journalists had a major scoop.

The story would only get bigger. As details of the forthcoming test were made official, media organisations clamoured to be included in the official party. They were flatly denied. That was never going to stop them. Norm Milne scoped out alternative observation posts that would not be in breach of the prohibited area. In June 1952, he wrote a memorandum to his news editor, outlining the options. He had been advised by Captain Morris that the newly gazetted restrictions imposed by the *Defence (Special Undertakings) Act* would not prevent journalists from using North Sandy

Island or Airlie Island, to the south-east of Barrow Island.[65] Morris went so far as to privately advise Milne that North Sandy Island, about 20 kilometres from the coast, would offer the best vantage point, provided the test actually occurred at Monte Bello and not, say, Rowley Shoals.[66] In the event, these potential observation points were not used. Instead, Milne was part of a group of journalists who established a rudimentary camp on the mainland, and waited out the delays on a high, windy redoubt until they could witness the blast. Although most accounts of the era placed this lookout at Mount Potter, in fact it was 12 kilometres to the south of that part of the Rough Ranges. Leonard Tyte believed that Mount Potter was the location at the time, and noted it in his diary. He said that the press being able to photograph Hurricane was 'completely unacceptable from our point of view since it could reveal the nature of the burst'.[67] Mushroom clouds are, however, impossible to hide. The British high commissioner in Australia lamented that options for stopping the journalists were limited. 'There is nothing by which the Hurricane panel are empowered to prevent this set up even if they so wished, short of difficult task of proving an intention to commit a felony.'[68] This possibility was at least contemplated – the High Commission considered informing the Prime Minister's Department of what the journos were up to 'with a view to clarifying whether the press are to be held vulnerable to prosecution if photographs are taken even from outside the prohibited area of what occurs within it'.[69]

A party of nine left Perth on 16 August, led again by *West Australian* reporter Jack Nicoll, bringing with them more than 6 tonnes of equipment and a determination to get images of the first British A-bomb. Norm Milne was with them, as were his *West Australian* journo colleagues Jim Cruthers and Peter Barnett and photographers Dick Long and RO Williams. Another photographer, Doug Burton from Perth's *Daily News*, was there too, along with driver Alec Laughton and mechanic Bob Warren. They were joined later by air by *Daily News* reporters Jack Coulter and Dan O'Sullivan and photographers Harold Rudinger and Bill Mangini (who was also a physicist). Pilot Jimmy Woods would fly the photos down to Perth after the explosion.[70] Chief among the gear they brought with them was state-of-the-art long-range photographic equipment.[71] The initial group drove a jeep, a 6-ton truck and a sedan car up the coast and initially stayed at the increasingly buzzing little town of Onslow before venturing forth to a shady billabong site, chopping away the scrub and spinifex and erecting tents.[72]

Once they had established camp at this tranquil, beautiful waterhole, a nearby lookout site was selected as the vantage point to view Hurricane, soon bestowed with the name 'Nick's Nob' to honour their fearless leader Nicoll. Although the journalists did not know exactly where the test would be held, their lookout would prove to be precisely 100 kilometres from the last anchorage of *Plym* at Main Beach, next to Trimouille Island. For two days they tramped up the hill. They carted nearly 2 tonnes of equipment, including a complete photographic darkroom.[73] The winds were close to unbearable and made rest difficult. The long-range cameras were mounted on heavy posts cut from timber at the billabong camp. Digging holes for the posts was impossible in the hard ground, so the journalists obtained some gelignite from Onslow and blasted out holes.[74] A cartoon drawn by Paul Rigby and only intended for the amusement of the group, not for publication, lampooned the hardships they faced. This raunchy image, which crudely depicts the sex starvation of the men, along with other features of the expedition such as the fierce winds, the difficulty sleeping and the riotous celebrations once the bomb was exploded, was titled 'Cactus Ad Infinitum (Operation Frustration)'.[75]

Today at the lookout, high on a hill with the local Aboriginal name of Bulyroona, the rock structure that provided the anchor point for the living quarters and dark room is still there. It is crumbling and tumbling, but the foundations of the two-room building can still be seen on the windy hilltop, surrounded by discarded bottles that attest to lengthy habitation by a group of media types. The letter 'P', painted on a rock a little away from the structure, marked the ablutions area. The group endured great physical hardship, not least having to deal with abundant spinifex, the spikes of which get into the skin unrelentingly, even through clothing. The steep climb, with rolling stones under foot, must have been nearly unbearable while carrying heavy gear. They did not know for sure until it happened that they would even see the bomb from their vantage point.

Then, on 3 October, they saw it at last. They did not have a perfect view of the Hurricane test, and for a while they believed that the bomb had been detonated on Flag Island or Hermite Island rather than aboard

Plym.[76] Milne's first story after the test said that the journalists 'heard a report like a clap of thunder, followed by a prolonged rumble like that of a train going through a tunnel. The air and ground shocks were sufficiently intense to cause slight pain in the ears'.[77] They had no official announcement to work with, and no time to lose once the bomb went up. Fortunately they had at their disposal a clever way to ensure that their stories could get back to Perth quickly. A fully official 'post office' on the back of a truck was set up near the billabong camp to enable the media to send their stories by telegraph.[78] It was called the Mount Potter Post Office even though it was not at Mount Potter, and the name reinforced ongoing confusion about the actual location of the journalists' lookout. This unique post office did not sell a single stamp or handle a single letter. Instead, when it opened on 28 August 1952, its sole task was to serve the journalists and photographers reporting on Operation Hurricane. Two telegraphists, Edwin Rodgers and Roy Buchanan, were flown from Perth for the task.[79] The airstrip at Mardie, 40 kilometres from the lookout, had a chartered Anson aircraft standing by for Jimmy Woods to ferry the pictures of the explosion to Perth, where they would be processed to provide imagery for all media, around Australia and the world.[80] A telephone link between the lookout and the post office was set up.[81] Within two minutes of the explosion, the telegraphists were hard at work, transmitting the words provided by the journalists. At 6.59 pm on 5 October 1952, Mount Potter Post Office closed forever, the staff headed back to Perth and the truck was recovered and sent to other duties.[82] The story of the decade and accompanying dramatic photos blazed across the front pages of newspapers around the world.

As the Nick's Nob crew watched impatiently, buzzed by flies and passing RAAF pilots, Operation Hurricane built to a crescendo out at the islands. The scientists and military personnel laboured mightily, awaiting the moment when William Penney, intensively studying the meteorological data, would deem that the weather was safe for an atomic test. Only on his command could the countdown be initiated, and only if the weather was amenable would he give the order.

Meanwhile, HMS *Tracker* was readied to deal with any medical emergencies once Hurricane was detonated. A British air force officer, Wing Commander Geoffrey Dhenin, was stationed on board *Tracker*. (Dhenin would in 1953 pilot the RAF Canberra bomber that flew into the Totem 1 cloud six minutes after detonation of a British bomb at Emu Field, South Australia.) After Hurricane, 91 recovery and salvage parties were put ashore and brought back to *Tracker*, according to the AWRE radiation hazard division's Alec Walkling.[83] Wing Commander Dhenin was attached to the navy for temporary duty as RAF medical observer, becoming ship's doctor from May until 22 October 1952. His diary shows insights into the medical side of the test operation, along with a frank disdain for many navy practices. On 17 June 1952, he noted the large number of personnel suffering from skin conditions and asked the rhetorical question: 'Are men given an examination for fitness for overseas service in the Navy? A large number slipped through if such an examination is routine'.[84] A litany of complaints about unserviceable equipment and poor medical processes appears throughout his diary, as Dhenin endured his placement. His entry from 4 August is more laconic than despairing, however: 'Sailed for the Monte Bello Islands. No cases of venereal disease yet reported'.[85] Upon encountering huge quantities of flies at the islands, and an uptick in cases of gastroenteritis that he thought was related, he commented that a 'large scale fly-destruction campaign is required. The Naval D.D.T. solution is so weak as to be almost useless'.[86] British crews never did get used to Australian flies, at any of the atomic test sites, and more than once called plaintively for their total eradication.

One of Dhenin's tasks was to prepare for radiation-related issues. On 20 August, he set out his plan should casualties occur on contaminated ground. He was to be in constant radio contact with all landing parties to advise them promptly on medical issues and offer them options depending on their needs. These ranged from first aid by leader of the party to sending a medic to them by helicopter.[87] He oversaw a range of medical briefings, and arranged for all crew to have their blood tested. His final entry before the bomb went up was simply: 'All blood counts are now finished. We await the event'.[88] His entry for 3 October was somewhat lacking in detail, but no doubt he was busy seeing crew members that day. He wrote: 'The weapon was exploded this morning at 9.30 a.m. [actually 7.59 am] There were no casualties. TRACKER at once took over the functions of Health Control as already described'.[89] On his last day on board, 22 October 1952, he handed over to RAF Squadron Leader R Mortimer, letting him know that there had been no radiation casualties or accidents in the contaminated areas.[90] He returned to the UK by air. Squadron Leader Mortimer continued the journal, noting that HMS *Tracker* left the Monte Bello Islands at dusk on 31 October. 'The film shown in the wardroom that night was called, appropriately, "Outcast of the Islands".'[91]

The team leader of the AWRE radiation hazard division during Hurricane was Alec Walkling. Under his supervision was James Hole, a radiation scientist who had also been an RAF pilot during the war. Hole went on to lead health control in 1956 at Operation Mosaic as well. He instructed personnel on site in the correct use of film badges and dosimeters, equipment that he had had a role in developing after the war.[92] Hole and his team were responsible for issuing radiation film badges to everyone present at Hurricane, including those who were thought unlikely to come into contact with radiation. These badges were meant to be fastened at the waist using a safety pin for greatest efficacy, but were often worn incorrectly or not at all, a point on which he was closely questioned during the Royal Commission.[93] Hole was based on *Tracker*, where he created a map that recorded the radiation survey results after the Hurricane blast and conditions on various parts of the islands. He used this map to decide on the correct allocation of film badges and dosimeters.[94]

William Penney arrived at the islands on Monday 22 September, having flown into Onslow secretly and been transferred to HMAS *Hawkesbury* for the trip out to sea. He was travelling with the Canadian defence scientist Omond Solandt, and two of the Australian observers, Alan Butement and Ernest Titterton, Americans having been 'pointedly not invited'.[95] Penney had met and worked with Solandt during the late 1940s, when Canada had been seriously investigated as the site for British tests.[96] McEnhill said that Penney's presence on board *Campania* 'greatly improved the morale and spirit of the entire company. His arrival was as a breath of fresh air on a somewhat jaded and slightly stale force'.[97] Penney immediately got to work on the meteorological aspects of the coming trial, including some pessimistic forecasts that were circulating through scientific ranks that fallout on the mainland was likely.[98] 'On examining the situation Dr. Penney decided that the contamination issue might well be an exaggeration of the true hazard but that it would be possible to avoid it by judicious use of upper wind direction for the trial.'[99] He pushed forward with an early October trial date.

Planning for the lead-up to detonation was divided into phases, each named after planets. Phase Mercury began the day before detonation and included final preparations on the land sites, up to the point of evacuation of personnel.[100] At 6.30 pm on the evening before the blast, Phase Venus was called, and ships moved to their safe positions. At half-past midnight on 3 October, Phase Mars was initiated. The radioactive components of the weapon were loaded, a process that was completed by 2 am.[101] Phase Jupiter involved the insertion of electric exploder units into the weapon assembly, completed by 4 am. At that point, Phase Neptune began, testing the telemetry. As McEnhill recounts, at '0600 [Charles] Adams aroused the sleeping staff at H.1 [Hermite Island] camp by booming out a message on the loud hailer informing them that it was in fact D-day'.[102] The men were able to wash and have a decent breakfast before the main event. Phase Saturn began at 6.15 am with crew on board *Plym* evacuating the ship and heading to HMS *Tracker*. *Plym* was left to its fate at 6.50 am, and Phase Uranus began. Vessels moved to their safe positions 16 kilometres east of Hermite Island and the weapon operation party finalised its preparations on the island. With unfortunate timing, some naval stewards who had been burning rubbish had accidentally set the tinder-dry spinifex on fire near the camp on Hermite Island and threatened to derail the whole operation as the fire spread rapidly.[103] Frantic fire-fighting efforts got the blaze under control in time, however, and Phase Boreas could begin. This phase involved final meteorological observations using a weather balloon sent aloft from Hermite Island to calculate the lower winds. When these observations came back as favourable, the all-clear was given. There was a brief delay to deal with last-minute problems with the ultra-high-speed Kerr Cell cameras, needed to capture images of the early stages of detonation, eliciting 'a rich stream of language' from the technicians trying to sort it out, then the countdown began.[104] The complicated, expensive and often difficult build-up was about to end. The crew on board the many ships stationed at the islands waited impatiently for that moment. Soon they would be able to witness the fruits of their labours.

CHAPTER 4

PLYM NO MORE

> The Navy remembers its great ships, but until today few had heard of *Plym*. She never sank a sub, she'd never even seen one. No record to speak of, no heroic action to remember. But if she goes to kingdom come like this, perhaps she will prove the greatest of them all.[1]
>
> *THIS LITTLE SHIP*, ADRIAN COOPER AND RON OSBORN (DIRECTORS), UK ATOMIC ENERGY AUTHORITY, 1953

> So down in stages of five seconds counted Mr. Maddock until five, four, three, two, one, zero. At this instant through minute cracks in the front wall of the building the observant could see a brief spark-like glint from the detonating bomb ... [The scientific and military personnel] could see the vast crisply outlined water column rising rapidly in the stillness. This was the most impressive feature of these early seconds – the vastness of the upheaval, its splendid symmetry, and the stillness.[2]
>
> JJ McENHILL, SCIENTIFIC WITNESS TO OPERATION HURRICANE FROM HERMITE ISLAND, MONTE BELLO, 1952

Operation Hurricane has gone down in history as the moment when the world's third atomic power triumphantly emerged, more than seven years after the Manhattan Project bombs were deployed to such devastating effect in Japan. Britain was a fair distance behind both the US and the USSR, and chronically underfunded and under-resourced, but eager to assert what was left of its great power status. The Manhattan Project built on fundamental physics research done earlier in Britain and elsewhere in

Europe to develop and test the world's first atomic weapons. It was a huge scientific and engineering undertaking, in which British scientists, including William Penney, played a key role following delicate negotiations between Winston Churchill and the US wartime president Franklin Roosevelt. In August 1945, these weapons were used against civilian populations in the Japanese cities of Hiroshima and Nagasaki, causing at least 185 000 deaths.

The story of Operation Hurricane, though, is one of compromises, uncertainties, errors and insufficient preparedness. The 1984 Kerr Report was scathing, saying that Hurricane was 'poorly documented and essential details may never have been recorded'.[3] The rush to get this weapon tested put a huge strain on those charged with designing, building and detonating William Penney's cumbersome Blue Danube weapon. The lack of critical appraisal of the project by the Australian government is also one of Hurricane's defining features. Canberra said yes, then the wheels turned quickly to add the Monte Bello Islands to the short list of places that had been desecrated by an atomic weapon.

The Hurricane device was a big bomb: about 35 per cent bigger in fire power than the 'Little Boy' bomb dropped on Hiroshima. The detonation created a dramatic cloud filled with radioactive particles that lingered over the Australian mainland and was tracked across the continent to cities along the Queensland coast. The British bomb makers heaved a sigh of relief, since the success of Operation Hurricane had been far from assured. Captain Pat Cooper of the Royal Navy mused later that 'the margin between success and failure was dangerously narrow'.[4] While the Blue Danube device was similar in concept to the bomb dropped on Nagasaki three days after the destruction of Hiroshima, it was a unique and untested British design. A nuclear catastrophe that might have killed people at Monte Bello was among the range of possibilities on that spring day in 1952. The moment of detonation when it came, though, was exultant.

The UK Atomic Energy Authority produced a moody, poignant short film in tribute to HMS *Plym*, titled *This Little Ship*. In it, the brief life of *Plym* was recounted, structured using the countdown of the last eight and a half minutes of the ship's existence. This overtly patriotic film, narrated by the actor Jack Hawkins, who was best known for portraying military men, was in some ways a paean to the old style of conflict for which *Plym* had been built, as exemplified by her service in the Second World War. Now she was being sacrificed in the name of a new kind of war altogether. In dramatic style, Hawkins said that war still had to be fought, and this time the weapons were the 'inheritance of men who, by the contrivance of their brains, had learned how to cut the bonds that hold the very substance of the Earth together'.[5] As the last crew left, having completed their tasks, Plym was 'delivered into the hands of science'.[6] The final countdown to 'NOW' is genuinely moving, knowing that *Plym* is about to be violently dismembered.

Another, rather less emotional, film was produced by the UK Ministry of Supply. *Operation Hurricane* showed the work involved in preparing for the Hurricane test, and highlighted British technical expertise. This production was put together by the Crown Film Unit. The film described the islands as 'a wilderness of saltbush and spinifex', and depicted the sheer physical effort that had been expended to turn it into a nuclear test site.[7] The film had a propaganda purpose, according to scholar Helen Hughes, in which it enjoyed great success, being shown in a number of countries, including the United States.[8] After all, Hurricane was primarily intended to prove to the Americans that Britain was sufficiently technically advanced to create this new kind of weapon. This film also featured the renowned countdown to the blast.

That countdown was delivered with clinical precision by Ieuan Maddock, whose role in Operation Hurricane earned him his nickname, 'The Count of Monte Bello'. A Welshman with a resonant voice and a solid background in physics and mathematics, he was recruited by William Penney to work at Fort Halstead (and later Aldermaston) on instrument development.[9] He was assistant director of telemetry at Hurricane. Aside from using his distinctive voice to count down to firing the weapon, he also worked on a wide range of measurements such as the implosion time of the bomb, firing circuit performance, high-speed cameras and associated equipment, and accurate timing and recording of the firing.[10] Until his death in 1988, Maddock was still receiving Christmas cards addressed to The Count of Monte Bello.[11] He would continue his association with the British tests by becoming the scientific superintendent for Operation Mosaic.

The Hurricane device was detonated on 3 October 1952 at 7.59.24 am. The five Royal Navy ships and several Australian ships were grouped to the south of the archipelago, their crews easily able to witness the blast. Other Australian ships that had been at the islands in the lead-up had moved away, some to Onslow harbour, some out to sea. Planning for the operation had specified that no ship could be closer than 10 miles (18 kilometres) to ground zero, and indeed most ships were at least 32 kilometres away.[12] The personnel who were closest to the blast were on land, at the command centre on Hermite Island, only 11 kilometres to the south-south-west of the target vessel. HMS *Plym* was torn apart in the 25-kiloton blast, and the seabed beneath it was deeply scarred. One eyewitness aboard *Campania*, the Royal Navy meteorological rating Michael Stephens, described it as 'a horribly dirty bomb'.[13] The huge rising column of greyish, contaminated water was an imposing sight, and the following spectacle of thousands of dead sea creatures rising to the surface was a vision of horror. The blast burst a bubble. After the frantic effort to be ready in time, both the scientific and military personnel at Monte Bello celebrated with raucous parties and prodigious consumption of alcohol.[14]

Young Dorrie Wally (also spelled Walley) was born at the huge Mardie Station on the mainland and lived there until the 1970s. Upon her passing in June 2022, Mrs Wally was considered to be one of the last Elders to be fluent in the Martuthunira (also known as Marthudunera) language of the Pilbara region.[15] Mardie Station, covering some 225 000 hectares, was established in 1866 as a sheep station on the coast of Western Australia, near the mouth of the Fortescue River. Dorrie did domestic work there and at the time of Hurricane she was a teenager. Years later, when she was interviewed for an oral history, she identified the Aboriginal name of the Mardie Station land as Wirrawanti.[16] Mardie was the closest human settlement to the test, being some 90 kilometres from ground zero, compared with Onslow, which was about 145 kilometres away. On 3 October 1952, Dorrie and the other workers were told by the station boss that something big was about to happen and to find a vantage point. They climbed up onto a station water tank. From there she and her friends watched in awe, using glass bottles like telescopes. She said that it was 'like a mushroom, lot of flame in the bottom … Flame just went "boosh" … There was a mushroom on the top'.[17] The sight must have been as confusing as it was awe-inspiring.

Many others were awaiting the event, from a much greater distance. Like Winston Churchill, Frederick Lindemann felt some trepidation before the detonation, which was taking place only a year after Churchill's return to power. On 17 September 1952, he sent a note to Churchill outlining the public announcements that would be required to accompany Hurricane, suggesting wording for a statement that could be issued immediately after the successful Hurricane detonation: 'It is announced by the Admiralty on the authority of the Prime Minister that a British atomic weapon has been successfully exploded in the Montebello Islands'.[18] Lindemann advised Churchill that the statement needed to go out promptly, for two reasons:

> [u]nless the official statement is out within an hour of the explosion it will be forestalled by the Australian Press which has set up a post on the mainland from which the explosion cloud will be visible. Secondly, as the explosion will probably take place at 8a.m. Australian time, it will arrive at 1a.m. The Admiralty alone of the departments concerned has an information office always manned on a 24 hour basis.[19]

He noted that a similar statement would be made in Melbourne and in Washington.

With concern for his elderly friend likely to be sleeping at that hour, Lindemann closed his note with 'I take it you do not wish to be awakened to receive the announcement!'[20] Churchill handwrote a reply at the bottom of the note: 'Lord Cherwell, Make sure it has gone off before you issue the communique. Otherwise we might look silly. W.S.C. 20/9/52'. He needn't have worried. The device designed by Dr Penney detonated as planned and gave all who witnessed it a jolt of awe at the sheer power unleashed. Penney would be knighted before the end of October for his efforts.[21] The 1400-ton HMS *Plym* was dismembered, though not completely atomised. The cloud that formed was distorted by the massive quantity of sea water in the up-rushing column. Penney would later explain in a public address, broadcast simultaneously on the BBC Home Service and ABC radio in Australia on 7/8 November 1952,[22] why it did not have a classic mushroom shape, saying it was due to strong winds blowing in different directions at different heights.

'The cloud was pulled into a gigantic spiral shape which, when seen from the ships and from the Mainland, appeared rather like a letter "Z", rapidly moving northwards away from the islands and from the Mainland.'[23]

Robert Menzies made a point of saying in his announcement following the detonation that he had been unaware of the timing of the blast until after it had happened, underlining the arms-length distance of his government from the test. His private secretary, Geoffrey Yeend, informed him only three or four minutes after the news was received at Navy Office in Melbourne. 'The radio flash conveying the news ... originated from the Royal Navy Flagship H.M.S. "CAMPANIA" at [10.30 am]. The radio flash contained the single code word which, it had been arranged, would indicate that the explosion had occurred successfully.'[24]

William Penney's top-secret director's report on Operation Hurricane indicated that anyone on land in close proximity to an atomic bomb explosion on board a ship would receive a very high, probably lethal, dose of radiation.[25] This was the kind of information needed to plan for an attack on a port. The report specified that the lethal dose rate was 400 roentgens, and this measure is accepted today as the 'LD50' threshold, in which 50 per cent of people exposed to this amount of radiation would die within 30 days.[26] The Monte Bello test showed that the risk was greater for people near a water-based explosion compared to an airburst. A further survey conducted by the AWRE in November 1953, 405 days after Operation Hurricane, found that the land areas surrounding the blast were still highly radioactive.[27] Indeed, despite 40 centimetres of rain and a cyclone in early 1953, 'there was no apparent evidence of the accelerated decay of fission products due to weathering'.[28]

Operation Hurricane was driven forward largely by geopolitical factors, with little in the way of concerns about safety or indeed much interest in anything other than the performance of Penney's Blue Danube design and the base surge experiment. Little effort was put into understanding fallout from the blast. The only attempt to measure fallout was by collecting and analysing water from roofs at several locations.[29] A roof about 500 miles (804 kilometres) from *Plym*, probably at Broome according to the Royal Commission, was treated with special paint and runoff water filtered to collect suspended particles. The filters were analysed to determine any radioactivity. As the Royal Commission recounts, similar filters were installed in Rockhampton, Cairns and Brisbane on Australia's east coast, and the capital of Fiji, Suva, in the Pacific Ocean.[30] 'No results were given for Broome, but activity in the filter from Rockhampton was about 200 times background and that from Brisbane about 10 times background. No activity was found in the filters from Cairns and Suva.'[31]

HMS *Plym* had been anchored in water approximately 12 metres deep for Penney's base surge experiment. The scientists were principally concerned with investigating how the weapon functioned and the contamination, blast and heat effects of the weapon.[32] When planning for Hurricane, the British had limited access to information about American tests, but saw some correlation with the second American test at Bikini Atoll, the Baker test, which was detonated much deeper under water. As the AWRE noted in its planning, the American bomb

> was exploded fairly well down in 200 ft. of water. About 150,000 tons of water and also a few thousand tons of calcareous sand were thrown in the air. The maximum height to which the water column rose was 5000 ft. The air blast was about equal to that of 5000 tons of TNT.[33]

Upon detonation, Hurricane did behave in somewhat the same manner as the Bikini Atoll test, according to Penney,[34] although the Hurricane cloud rose twice as high, climbing to over 3 kilometres.

An atomic explosion of this kind, in water, behaves differently to those on land. AWRE scientist David Stevenson, in his statement to the Royal Commission, explained that Operation Hurricane was very different

> in that the blast was cooled and the dust cloud mixed with a vast quantity of seawater so that much more of the soluble and adherent contamination fell to the ground near the site of the event. Clothing from this trial was more heavily contaminated with a more tenacious type of material.[35]

For Hurricane, decontamination was in the hands of the Royal Navy. Nevertheless, Stevenson became involved with the clothing of personnel.

He told the Royal Commission that the most vulnerable areas for personnel were around the neck and head, since the protective clothing worn by re-entry parties did not have hoods. This deficiency was rectified for Operation Totem at Emu Field the following year.[36] Stevenson began studying the characteristics of contaminated clothing from Hurricane, and used this knowledge later when he was present at RAAF Base Pearce for Operation Mosaic in 1956, in charge of aircraft decontamination. HMS *Tracker* crew member Reginald Garbett spent three months at the Monte Bello Islands, where he was part of a team responsible for destroying the clothing that had been worn ashore. He reported many years later to the Royal Commission that '[n]one of us performing this operation of handling and destroying the clothing were given any protective equipment to carry out this operation'.[37]

Winston Churchill made a statement about Operation Hurricane in the House of Commons on Thursday 23 October 1952, nearly three weeks after the test. The Australian government had been informed of Churchill's statement and shown a draft the day before.[38] In scripted words that had been carefully plotted before the Commons session, Churchill said that thousands of tonnes of water and of mud and rock from the sea bottom were thrown thousands of feet into the air, causing a tidal wave. 'The effects of blast and radio-active contamination extended over a wide area and H.M.S. "Plym" was vaporised except for some red hot fragments which were scattered over one of the islands and started fires in the dry vegetation.'[39]

The statement continued:

> Very soon after the explosion two naval officers undertook the dangerous task of flying helicopters over the heavily contaminated lagoon where 'Plym' had lain. This was in order to take samples of the water so that its radio-activity could be measured. After a longer interval, scientists and Service personnel in protective clothing entered the contaminated area to examine the effect and to recover records.[40]

While declining to make any disclosures about the technical details of the blast, Churchill offered assurances that no person or animal had been harmed; apart from 'some local rats which were killed, no mammals were seen in the affected area and such birds as there were had mostly been frightened away by the earlier preparations'.[41] After his statement, when questioned by shadow Cabinet member and former defence minister in the Attlee Labour government, Emanuel Shinwell, about whether the US had taken an interest in the test and been supplied with information about it, Churchill made a jab at American recalcitrance represented by the McMahon Act, saying, 'I do not doubt that it will lead to a much closer American interchange of information than has hitherto taken place – than has taken place in the last two years – but I do not wish to make any statement on the subject'.[42] Under further questioning, Churchill made it clear that the success of Operation Hurricane meant that the door slammed shut in 1946 had been cracked ajar, and Britain intended to go back to work with the Americans one day. When Churchill rather grudgingly acknowledged that the Attlee government had laid the groundwork, a hard-left Labour member who had a role in founding the Campaign for Nuclear Disarmament, Sydney Silverman, spoke up. He said: 'to some of us it is no comfort at all to realise that both the major parties in the State are equally responsible for this colossal folly'.[43]

The man of the moment was trial director and weapon inventor William Penney. In his November radio broadcast, Penney talked proudly about the bomb he designed.[44] His broadcast was timed for after Churchill's address to the Commons, and he referenced Churchill's words. He gave some insights into how the operation took place, saying that the needs of civil defence drove the maritime design of the trial, alluding to the risk posed by ship-borne nuclear attack. Hoisting a device onto a tower would have been the easiest option, he said, but this was not done because of 'weighty considerations' that meant simulating an atomic attack in a port took precedence.[45] To gather the required data, *Plym* was wired up to be a 'scientific transmitting station, sending out by radio a vast number of measurements about the nuclear explosion before the equipment was destroyed'.[46] He described the distinctive cloud:

> The sight before our eyes was terrifying – a great, greyish black cloud being hurled thousands of feet into the air and increasing in size with astonishing rapidity. A great sandstorm suddenly sprang up over the islands. It seemed ages before we heard the bang but, in fact, it was only a minute. Somewhat to our surprise, a second bang –

> at least as loud as the first – followed a few seconds later. At the same time we felt a peculiar sensation in our ears such as one has in an aircraft losing height rapidly. We were feeling the suction, or reduced pressure, which always follows a blast wave.[47]

Penney admitted to some anxiety before the test, particularly concerning meteorological conditions, for which two consecutive days with strong winds were needed.

> We got the strong winds all right! The winds were so strong that we were not sure that the boats could get the scientists all round the islands to make the final adjustments of their instruments. However, as always, the Navy and the Marines rose to the occasion and the work was completed with nothing worse than a lot of wet shirts.[48]

The two bangs Penney and others experienced were accounted for by the fact that the direct sound wave from the blast was reflected back when the sound bounced off a layer of warm air over 3 kilometres in the sky. Not mentioned by Penney, but noted many times by other eyewitnesses, were the effects of the blast on local sea life. RAAF Flying Officer Keith Peck, who was part of a joint services team that took radiation measurements on Trimouille Island after Hurricane, reported seeing 'very large lobsters, small lobsters, sharks, all sorts of marine fish, that were washed up dead, and if you approached those on the beach, then you would get a reading, sometimes a very high reading'.[49] The abundant, and in some cases unique, marine life of the Monte Bellos was an immediate casualty.

Henry Carter had served in the Royal Navy during the Second World War, eventually leaving the service in 1948. He was recalled as a Fleet Reservist at the end of 1951, and did some refresher courses. He was then drafted to HMS *Plym*. The vessel was at that point in Chatham dockyard in Kent, being extensively fitted out for an unknown mission. As Carter related in his submission to the Royal Commission, there were 'many stories as to the purpose and use of HMS *Plym* but the truth was not known until we were at sea on the way to Monte Bello Islands from Freemantle [sic]'.[50] His task was to be among the last to leave the ship, with one other seaman.

They were required to take the scientists off the ship and, if the bomb did not detonate, bring them back. After escorting the scientists off *Plym*, Carter and his colleague drove to where a small power boat and several other service personnel were waiting. They travelled a zig-zag course in the boat and eventually sheltered behind a ledge on Hermite Island. Carter estimated in his submission that they were about 2 miles from *Plym*, amended to 5 or 6 miles (8 to 9.5 kilometres) during his testimony.[51] Word came that the bomb was about to detonate. The men pulled a heavy canvas tarpaulin over themselves and draped their heads in jungle green towels. Carter, who was dressed in shorts and a pair of shoes, pushed his knuckles into his eye sockets.

> At zero there was a blinding electric blue light of an intensity I had not seen before or since. I pressed my hands hard to my eyes, then I realised I could see the bones of my hands. It seemed that this light was passing through the tarpaulin and towel for about 10 or 12 seconds and there seemed to be two surges and two detonations with a continued rumbling and boiling sensation.[52]

He said he felt like his body was at first compressed, then billowed like a balloon.

He was checked for exposure on board *Narvik*, and his pay book was stamped 'Nil Roentgens', referencing the then-current unit of measurement of some forms of radioactivity. But the following day, when he was on board HMS *Campania*, he became ill with what seemed to be a chest cold, an illness that recurred on the journey home. Several days after the Hurricane detonation he flew in a helicopter over the scene 'and saw that everything was black and burnt including the sand and rocks right up to the ledge we had been sheltering behind. I believe that the holes in the tarpaulin were caused in the same way as the blackening of the area'.[53] Upon his return to England, Carter felt unable to pick up his five-month-old son and avoided him. That son would die at age 24 from multiple cancers. His daughter, who was conceived after Carter's experiences at Monte Bello, suffered from severe eczema. At the time of his submission, he had had a rash on his back for 32 years and had undergone cataract operations. 'No doctor will tell me what is wrong with me but I have lived on a daily intake of drugs for 14–15 years and I have attempted to obtain a copy of my medical file which I have been told has been "stripped".'[54] Later, when questioned at the Royal Commission about his submission, Carter denied that he had been told that his file had been 'stripped', but just that he had come to that conclusion himself.[55] He gave evidence that he had been 'ultra-fit' when recalled for service in December of 1951, but had deteriorated into ill-health after his trip to the Monte Bellos to the point where he was unable to do physical work.[56]

RAAF Squadron Leader Andrew Thomas led the first party to Trimouille Island after the Hurricane blast. Thomas's extensive experience with radiation measurement meant that he was the only Australian in the landing party.[57] They spent about two hours on the island, wearing full protective gear – suits with elastic at the ankles and rubber boots, a sweat cloth around the neck, a dustproof hood and a respirator.[58] In 43-degree Celsius heat, the party drove towards the part of the island adjacent to where *Plym* had been anchored. As they did so, their radiation monitors reacted. 'We finally reached a point where there was a piece of metal from Plym. The meter that I had, had a maximum reading of 30 [roentgens] per hour and as we approached the piece of Plym the needle went to full scale. Immediately I told the driver to turn around and get away from it quickly.'[59] Thomas and his crew were involved in collecting samples of grass, seaweed, fish and shells. They also attempted to find animals on the island to test, without much success. In addition, they carried out tests such as measuring the cracks in brickwork and concrete structures that had been built before the explosion.[60] Thomas noted once he returned to the health ship, HMS *Tracker*, that he had lost more than 3.5 kilograms in two hours from perspiration. Captain Cooper noted in his later report that the average loss of weight of men working in the aftermath of Hurricane was 7 pounds (3.1 kilograms), 'although 14 or 16 lbs. were occasionally lost – usually by men of ample proportions who could afford it'.[61] (Penney also witnessed the physical toll exacted by going into the contaminated area, although he brushed it aside in his broadcast with the airy words: 'on return to the health control ship, a few long drinks of water, some salt tablets and a meal with lots of tea, quickly restored the loss of weight and nobody felt any the worse'.[62])

Andrew Thomas was asked by Admiral Torlesse to monitor radiation during the salvage of a landing craft vessel that had sunk off Daisy Island during a storm more than a week after the Hurricane blast. The landing craft auxiliary (LCA), a small flat-bottomed boat, sank off Daisy Island, on the western side of the parting pool. He was to board HMAS *Koala*, which was suitably equipped for salvage operations. *Koala* duly hauled the vessel to the surface and Thomas was able to monitor it. He said,

> the monitoring revealed that it was heavily contaminated with radioactive mud, sand, seaweed and the like. *Tracker*'s commanding officer decided that he did not want the boat aboard at all and advised *Koala* to dump it in deep water near Trimouille Island as soon as possible. Subsequently it became apparent to the British that HMAS *Koala*'s equipment and parts of the ship may have become contaminated as a result of this recovery exercise.[63]

This included the carpet in front of the captain's desk in his cabin, where officers and sailors had come to talk to him while wearing contaminated footwear.[64] Thomas was the only person on board with experience with decontamination, and he was expected to make the vessel safe. He endeavoured to do so with the help of seamen who were rapidly trained in the use of scrubbing brushes and Teepol detergent, who pitched in and followed his instructions during the 48-hour process. Much of the moveable material that was contaminated was simply dumped overboard, Thomas concluding that it couldn't worsen the already-contaminated ocean environment.[65] Over the side went the captain's carpet. The ship's cat was monitored, and fortunately found to be clear of radioactivity. Thomas said, 'I was concerned because he had probably licked himself clean and I had no way of knowing what contamination he had taken on his tongue and ingested. However I understand that he lived a long time'.[66]

On its way back to Fremantle after Hurricane, HMS *Zeebrugge* quietly slipped steel drums of waste from the test over the side and into deep waters. Most of the waste was liquid, but there was some solid waste and contaminated fittings from the onboard laboratories.[67] Graham Mabbutt was an acting petty officer on the vessel, and at the Royal Commission he recounted the slapdash way the task was conducted. He said that there were 20 to 30 drums, and to his certain knowledge up to eight of them were seeping badly. The mechanism for throwing the drums into the sea was not working properly, so Mabbutt had to provide manual assistance.

> I remember getting splashed over the arms and legs by the seeping liquids from these drums. I can clearly remember a scientist who was observing these operations saying when he saw me being splashed 'one day you may live to regret that'. This raised a laugh among the ratings present as it was taken to mean that it might affect my ability to have children.[68]

Upon his return with the ship to England, Mabbutt was part of a team that stripped the paint off the bottom of the ship, working bare to the waist.[69] Soon after he began to suffer from psoriasis on his arms 'and a nervous state which is quite foreign to my nature',[70] although his health improved once he left the navy and joined the police force. Years later, at the Royal Commission, the team leader of the AWRE radiation hazard division, Alec Walkling, would deny that there was an issue, and would maintain that the dumping of contaminated waste from *Zeebrugge* was unproblematic. Walkling described how the drums that wouldn't sink were shot with a rifle.[71] He claimed that the waste contained 'such a low level of activity that it did not warrant the procedure laid down in Hurricane trial orders which was that it should be transferred to Tracker for disposal'.[72] The media picked up on Walkling's testimony in London, quoting him as saying, 'The possibility of any hazard from such contamination was absolutely negligible ... there was no danger'.[73]

The British set up some 'target response' experiments as part of Hurricane, to test what would happen to equipment in the path of an atomic blast. Anderson shelters – Second World War bomb shelters made from corrugated iron, of the type installed in many British backyards during the war – were set at a variety of distances from the blast. The shelters, covered with sandbags, were placed at 420, 466, 548, 841 and 1033 metres from ground zero.[74] Those at the three closest locations suffered damage, including demolition of blast walls, removal of sandbag covering and displacement of corrugated iron. Anyone hoping that old-style bomb shelters might provide safety in the event of a nuclear attack would be dismayed by the results of these experiments. The AWRE report estimated that half of personnel would survive blast damage in shelters at 420 metres and about 90 per cent at 466 metres, and that fatalities would mainly be due to secondary blast effects such as flying debris.[75] The front part of the shelter was the most dangerous area, due to the collapse of the blast wall into the shelter. These calculations were rather academic, however, because at such distances, no one would survive the effects of the gamma flash. Those at 841 metres and 1033 metres may have fared better.

Similar experiments were done with concrete cubicles, 3.6 metres cubed, with varying wall thicknesses and reinforcement. These structures outlasted the Anderson shelters, although the walls of some that were closer to the blast were caved in.[76] Measurements done in 'slit trenches' dug 2 metres deep into the earth showed the likelihood of lethal effects. These experiments indicated that people sheltering in trenches would receive a lethal dose at around 1200 metres from the blast. However, if the trench was covered in 30 centimetres of earth the dose was cut significantly, meaning that standing in a trench about 600 metres away could be survivable.[77]

While none of the scientific or military personnel suffered immediate ill-effects from the test, many would suffer a variety of health problems later. There was one casualty on site, however. The Royal Engineers would experience tragedy just after the Hurricane blast. Sapper Frank Furlong of 71 Field Squadron died on Hermite Island on 13 October when the bulldozer he was driving rolled over and crushed him.[78] His death was registered in Perth.[79] He was initially buried on the island with full naval honours, but the following year, when HMAS *Junee* came back to Monte Bello bringing scientists to conduct a radioactivity survey, it had on board two undertakers from Perth who came to exhume his body and take it to Onslow for reburial.[80]

HMS *Campania* and the rest of the Royal Navy fleet departed the Monte Bellos on 31 October, leaving behind a mess for the Australians to manage. Penney had advised that the site would need guarding for two or three months, a modest timeframe that he was questioned about years later at the Royal Commission and which he still held to be sufficient.[81] The main concern was for security, not radiation safety. A secret note from Captain Lloyd at AWRE to Vice Admiral Evans-Lombe at HUREX indicated that this was 'to prevent the possible collection of intelligence which is available on the site'.[82] Traces of the Hurricane weapon could be found in the form of fission products and unfissioned plutonium from the weapon, able to be analysed by a suitably trained person who could determine the nature of the nuclear fuel powering the British bomb. Every atomic weapon detonation leaves behind a signature. Based on Penney's advice to maintain a military presence, *Hawkesbury* remained at the islands 'principally for counter-espionage duties'.[83]

During this time, the Joint Services Training Unit (JSTU) was established by HUREX and the Australian Hurricane Panel to deal with the aftermath of Hurricane. Representatives from the three services were attached to the unit, and HMAS *Hawkesbury* was assigned to support it. The JSTU was headed by Arthur Andrews, a British-born RAN officer.[84] Andrews had spent a year at AWRE in 1950, studying atomic, chemical and biological weaponry. He was not present for the actual test, but flew to Onslow directly afterwards, and was taken across to the islands by ship with a group of Australian service personnel.[85] Because he had had some rudimentary training in physics and atomic weaponry, he was charged with teaching basic principles to the group. Andrews would later be involved in Operation Mosaic, and with trials at Maralinga, and retired from the RAN as a lieutenant commander in 1962. His statement to the Royal Commission did not mention his work with the JSTU, although he was questioned about it. He gave evidence that he had started the ABC School, standing for Atomic, Biological and Chemical Warfare (also known as the Nuclear, Biological and Chemical Defence School), as part of the naval training school at the naval base HMAS *Penguin* in Balmoral in Sydney, and claimed to be the only officer in the RAN who had knowledge and experience in this area. Some evidence was presented that challenged that assertion. The JSTU group set up camp on South East Island, close to where *Plym* had been blown up, and used radiation monitors and other equipment to make radiation maps and carry out other training exercises. Andrews told the Royal Commission that one member of his party had disobeyed safety orders and had gone into the area contaminated by Hurricane to pick up souvenirs – 'he was dealt with in true naval fashion ... I sent him back to his depot'.[86] This comment refers to 25-year-old RAAF Flying Officer Keith Peck, who did not get on with Andrews. Peck and his fellow RAAF flying officer colleague JE Nicholls formed the view that Andrews knew little more than they did about the radiation-measuring equipment, and they knew nothing.[87] Peck and some of his colleagues opted to camp on South East Island instead of staying aboard *Hawkesbury*, another source of contention. In Peck's version of events, he was found, after several weeks of surveying Trimouille Island, to have ingested radioactive particles and he was medically evacuated from the island.[88]

HMAS *Hawkesbury*, which had been present for the operation, was handed the additional task of supporting the JSTU and assessing the clean-up required, relieved by some R&R in Fremantle in-between. Before heading south, *Hawkesbury* had taken scientific stores to Onslow, then come back to the test area, anchoring near *Campania*, about 11 kilometres south-east of Flag Island. The ship then had to shut down its evaporators for three days because of high levels of radioactivity in the sea water.[89] *Hawkesbury* headed to Fremantle for a couple of weeks, returning just as the British ships were leaving. Commander of *Hawkesbury*, Lieutenant Robert Scrivenor, was confronted with a large amount of equipment apparently abandoned. 'The majority of the equipment on Trimouille Island and North West Island is contaminated, in varying amounts, with radio-active material.'[90] A survey by the crew of *Hawkesbury* found nearly all of the materials used for target response studies, such as blown-apart Nissen huts and Anderson shelters, along with batteries, cables, generators, marquees, hoses, tanks, and much more.[91] The site was a mess. Scrivenor received orders from Western Australian naval headquarters to gather up as much easily portable valuable gear as possible and bring it to Fremantle, have a rest for two weeks then go back again for more.[92] During the fortnight that *Hawkesbury* was not there, the RAAF would operate patrols over the islands. The naval officer-in-charge West Australian Area cabled *Hawkesbury* to say that because Onslow locals were aware of *Hawkesbury*'s absence, 'full publicity to fact that area is still prohibited and highly dangerous will be necessary'.[93]

The debris was an unexpected headache for the navy, and the Naval Board in Melbourne was duly notified in a confidential note titled

'EQUIPMENT APPARENTLY ABANDONED'. 'As a considerable amount of this equipment is valuable and fairly portable it is intended that as much as is possible be recovered by "HAWKESBURY" and/or "KARANGI" as opportunity offers and held in store pending direction as to its disposal.'[94] The lack of information provided by Britain is a sub-text in this correspondence – '[n]o further information regarding this equipment is held locally with the exception that the "P.S.P." and the "Ships Tanks" and probably the 44-gallon drums were supplied from Australian sources'.[95] In common with the behaviour of the British test authorities at the two other Australian sites, Emu Field and Maralinga, they just walked away, leaving behind miscellaneous rubbish, at least some of it radioactive.

The Department of the Navy was worried about people visiting the site, despite warnings and publicity. Indeed, the legal position had been raised by the Department of Defence and the Attorney-General's Department as the Australian government started to realise the implications of highly toxic experiments on Australian territory. The concern was that 'warnings will not satisfactorily meet the legal position'.[96] Instead, keeping the affected areas under the prohibitive provisions of the *Defence (Special Undertakings) Act* was considered as a more prudent option. There were drawbacks, though, namely drawing attention to the actual places where the test was conducted, and removing restrictions on landing at places where valuable equipment was still stored, on Hermite and Alpha Islands.[97]

In any case, without the presence of *Hawkesbury*, even retaining prohibited areas could not guarantee safety, the Department of the Navy argued. In the meantime, the Department would await advice from the attorney-general. That advice does not appear to have been forthcoming, and in March 1953, the problem of the debris left behind at Monte Bello was still causing concern. Naval officer-in-charge West Australian Area wrote again to the Naval Board in Melbourne on 20 March 1953 to say that while *Hawkesbury* had recovered a number of items, the Navy had no idea what to do with them:

> These items have been accepted on 'Deposit Charge' pending further instructions as to their disposal. It has not been practicable, owing to a lack of trained personnel within the Command, to carry out a

> detailed survey of these items, but they appear to be mainly Ministry of Supply Stores, for which there is no known Naval service in the West Australian Area.[98]

Hawkesbury departed Monte Bello on 16 December, having first collected samples of seedlings, wheat, grass and other plants from both clean and contaminated areas to send to the Ministry of Supply for analysis.[99]

What the Australian officials did not know until June 1954 was that the site was far more contaminated than the British had first suggested. The original advice was that the site would remain 'hot' for about three years after Hurricane. However, in June 1954 the Admiralty advised the British government of a revised assessment, following a visit to the site by British scientists in November 1953. In a secret letter to the Admiralty, the secretary of the Prime Minister's Department, Allen Brown, wrote that 'the islands are still very contaminated and the conditions are not likely to change significantly over the next ten years'.[100] Further, Brown noted that 'it would be dangerous to remove any material or equipment from the areas without special precautions'.[101] This rather stiff letter hinted at the frustration the Australian authorities were starting to feel about the aftermath of the test, and the increasingly forlorn hope that Britain might step up and take some responsibility. The RAN had not been present at the Monte Bellos for 18 months and were not in a position to police the islands. 'It is desired to bring this to your notice so that the United Kingdom authorities might be fully aware of the position should they wish to consider what arrangements might be appropriately made by them on security aspects.'[102] No such additional security assistance was forthcoming.

Meanwhile, the British authorities had been exchanging secret notes with each other in the wake of Hurricane, concerned about the considerable risk to health to anyone who visited the area, particularly given *Hawkesbury*'s departure. Their main concerns, though, were mostly to do with liability. Controller of Atomic Energy Sir Frederick Morgan wrote to William Penney on 31 December 1952, worried that any harm might be sheeted home to them. He wrote, '[i]t is important that [the Australians] should not be able to lay blame for any accident on the British Government'.[103]

The media saw the aftermath of Operation Hurricane first-hand when a group of journalists flew over the site on 4 August 1953. The flight was organised by the *West Australian* newspaper, and TC Bray, editor of the Brisbane *Courier-Mail*, was part of the group. He reported 'Australia's mystery islands are still "hot" in the radio active sense' and that Trimouille Island had suffered the most from the 'devastating lateral blast'.[104] He saw that the infrastructure of roads, Nissen huts, masts, marker buoys and much more had been left intact, and speculated that the site was merely 'sleeping' until the British returned for more tests, maybe as early as 1954.[105] While it would take a bit longer, Bray wasn't too far out. Britain returned in 1956, and used some of the Hurricane infrastructure.

William Penney arrived home from Hurricane a hero. When he landed at Lyneham airport in Wiltshire, he was asked by the waiting journalists what his next moves would be. He replied, with a winning smile: 'I have to send in my report to the government. When I have done that I shall have a short holiday and I hope to play some golf'.[106] The official Ministry of Supply film of the first British atomic weapon was more sombre than its inventor. At the end of the film *Operation Hurricane*, the narrator posed some questions:

> how shall this newfound power be used? For good or evil, peace or war, progress or destruction? The answer doesn't lie with Britain alone. But we may have a greater voice in this great decision if we have the strength to defend ourselves and deter aggression. That was the meaning of Monte Bello.[107]

CHAPTER 5

AT THE CROSSROADS OF FUSION

> We must do it. It's the price we pay to sit at the top table.[1]
>
> WINSTON CHURCHILL ON THE DECISION TO BUILD A BRITISH H-BOMB

> Dr Evatt: Does the Monte Bello proposal cover a hydrogen experiment?
>
> Mr Beale: No. It is an atomic bomb test. It is one of series that has taken place and has nothing to do with the hydrogen bomb.[2]
>
> EXCHANGE IN THE AUSTRALIAN HOUSE OF REPRESENTATIVES, 15 FEBRUARY 1956

Hurricane was never repeated. The two fission weapons tested at Emu Field, South Australia, in 1953 (Operation Totem) used greatly altered variations of the Blue Danube design in a totally different environment. The Hurricane device itself could not be carried by the new RAF V bombers. Its detonation proved the viability of the design, before the UK took off in several different directions in succession, notably at Emu Field, trying to create a viable 'austerity' fission bomb with the eked-out form of plutonium that could be produced in the new reactors at the Calder Hall plant in Cumbria. These would form the backbone (in a form modified from the Emu Field test weapons) of the British nuclear armaments deployed to the RAF until the end of the 1950s. However, fission-only weapons were starting to look passé, just as Britain was demonstrating mastery of the technology. The country would soon face a crossroads as it contemplated its future defence strategy.

Here can be found another line of demarcation between the first Monte Bello test and the later ones in the Mosaic series. Both Hurricane and Totem were characterised by the absence of formalised oversight by the Australian government. However, in 1955, in the lead-up to Operation Mosaic and after two test series that were largely opaque to the Australian government, the Australian Atomic Weapons Tests Safety Committee (AWTSC) was established. The AWTSC was far from perfect, and was criticised extensively at the later Royal Commission for creating a barrier to the flow of information from Britain to Australia. Nevertheless, it did represent an attempt by the Australian government to hold the British test authorities to account in terms of test safety, including giving it the power to veto tests on safety grounds. The British were not overjoyed at this development, although they had no choice but to accept it, having staged Hurricane and Totem on essentially their own terms. In a secret telegram to the secretary of state for Commonwealth relations, Alec Douglas-Home, referencing the growing Australian concern about nuclear test safety issues, the British high commissioner to Australia, Stephen Holmes, said in March 1955, '[w]e gather that we should be given the opportunity of commenting on proposed composition of Committee before it was actually formed'.[3] Britain was keen to ensure that this new committee would not constitute any sort of threat to its plans.

Neither Hurricane nor Totem was subjected to the scrutiny and more sophisticated measurements that were applied to Mosaic, or the Maralinga tests Operation Buffalo (1956) and Operation Antler (1957). The two major series of tests at Maralinga, comprising a total of seven weapons, were focused on refinements to the Blue Danube design but also, most importantly, development and testing of the more compact Red Beard design. This new weapon would be significant in the development of the British H-bomb, as it was intended to provide the fission trigger for a thermonuclear (hydrogen bomb) detonation, as well as being a standalone weapon. Red Beard was deployed briefly to the RAF before, like all British weapons tested in Australia, it was made obsolete by US technology.

Buffalo in particular was conducted in less than optimal climatic conditions, but it was still subject to more rigorous safety standards than either of the first two series. Indeed, as the 1984 Kerr Report noted, '[a]fter reviewing criteria for the Hurricane and Totem series, in 1955 the AWTSC recorded their opinion that meteorological facilities at the previous trials were inadequate and at future tests "every means should be adopted to make meteorological information as complete as possible"'.[4] Hurricane and Totem stand out as especially neglectful of safety oversight; however, although Mosaic was subject to stricter conditions imposed by the Australian government, it was still a test program of immense risk, and only a few concessions were made to Australian concerns.

Those first two British atomic series were very much part of phase one of Britain's atomic ambitions. Emanuel Shinwell, Labour minister of defence during the Attlee government of the late 1940s and one of the insiders to the development of the bomb, was quoted by researcher Richard Gott in 1963 as saying 'we were not speaking in terms of atomic deterrents at that time ... we did not believe that the atomic bomb itself would prove an effective deterrent'.[5] When the UK decided to pursue the H-bomb in July 1954, and a few years later got back to work with the Americans on nuclear weapons design and testing, the initial work done at Monte Bello and Emu Field effectively was consigned to something of a cul-de-sac, except in a general sense that any atomic bomb test advances fundamental physical and engineering knowledge. Nuclear warfare was imminent, at least in the minds of the UK's top brass. Operation Totem at Emu Field had been the first real opportunity for them to determine what would actually happen in the event of this new kind of warfare, as that series had had a number of experiments requested by the military to assess the effects of atomic weapons on troops and materiel. Britain was trailing its nuclear weapons competitors, and in fact at the time 'Cabinet took its decision in July 1954 weapons scientists at Aldermaston were still a long way from knowing how to make thermonuclear weapons'.[6] That agenda item would move to the top of the list in 1954.

The nuclear landscape was changing rapidly. As Britain watched the US (in 1952), closely followed by the USSR (in 1953), test hydrogen weapons, the country could see titanic shifts in geopolitics, and a role for itself as the world was increasingly shaped by the threat of these new weapons. The Chiefs of Staff had to re-evaluate UK defence in light of the advent of the H-bomb and severe problems with resources, inter-service rivalry and differing views about how a future war might play out. As military historian Jim Gledhill put it, 'Britain's armed services were not only competing for limited resources but also struggling to define their respective roles in the nuclear age'.[7] The H-bomb made more of a profound change to British defence policy than even the A-bomb, but the development of British thermonuclear technology was always a political not a military initiative, part of the ongoing quest to bring Britain and the US back together.[8]

Britain would need to answer the rise of the H-bomb, but what was the best course, particularly as it had already devoted tens of millions of pounds to creating a fission weapon? A briefing document created for the Air Defence Committee Working Party by EC Williams in May 1954 put the case in stark terms, saying that once thermonuclear weapons were produced by the US and USSR in sufficient numbers for global annihilation,

> further superiority in numbers has no meaning. This has the important consequence that we in the United Kingdom can, if we so wish, provide an independent deterrent of equal magnitude. By so doing, we gain the maintenance of our rating as a major power, prestige as the focus of the Commonwealth ... and so on: but we invite special attention from the U.S.S.R. as a threat of magnitude comparable to that of the U.S.A.[9]

Williams set out a horrifying scenario in which, in the event of wartime use of hydrogen weapons against Britain, the country would no longer be tenable or be able to sustain its population. 'There is little that the U.K. can do about avoiding these consequences of such a war: our whole energies must be bent to see that war does not happen.'[10] By becoming the third H-bomb-armed power, Britain saw itself as neutralising the threat of the two superpowers, creating a 'thermo-nuclear stalemate, in which, in this respect, superiority in numbers has no meaning, and equality is re-introduced'.[11]

While Britain had floated the idea of developing a hydrogen weapon well before it tested its first fission weapon, this was not the main focus until after Operation Totem. A question asked in the House of Commons by Labour member Albert Blackburn on 22 March 1950 raised the issue of America's pursuit of the H-bomb, which had been announced by President Harry Truman on 31 January 1950, nearly two years before that country was able to test such a device.[12] Blackburn asked the secretary of state for foreign affairs, the Labour government's Ernest Bevin, 'what proposals he has received from the Government of the United States of America in relation to atomic energy and the hydrogen bomb project; and what arrangements he has made for consultation with President Truman'.[13] Bevin responded, 'I have received no new proposals from the United States Government in relation to atomic energy and the hydrogen bomb project', but did not mention anything about the UK's ambitions in this area.[14]

However, the Ministry of Defence had carefully prepared for Blackburn's question, including an answer on the hydrogen bomb part, sharing it with Frank How in the Ministry of Supply on 8 March 1950:

(1) That we have no official detailed knowledge regarding the development of the hydrogen bomb though its broad principles will be known to any competent nuclear physicist.
(2) That for a long time to come we shall have quite enough on our plate in connection with the development of conventional atomic weapons.
(3) That, in any case, we should not wish to make any public statement as to our intentions in the matter.[15]

This part of the answer was not given in parliament, but provides an insight into the thinking at the time, 18 months before Churchill returned to power and more than four years before the Churchill government made the decision to pursue the H-bomb. Hydrogen bombs were exercising the minds of British politicians, but there were no plans at that stage to produce a British version.

These weapons were terrifying – a thousand times more powerful than the fission weapons that were fast looking outdated. Churchill was appalled by their destructive potential, but also keen to ensure that Britain could hold the balance of power between the two H-bomb-armed superpowers, equally concerned about American proliferation as he was about that of the Soviets. To achieve this equilibrium, he decided to tilt at the fusion windmill. In doing so, he fundamentally changed the nature of the tests being carried out in Australia. On 14 April 1954, Churchill invited his Cabinet to begin considering the prospect of a British hydrogen weapon.[16] On 16 June, the UK Defence Policy Committee agreed to pursue the production of the thermonuclear bomb, and on 26 July Cabinet approved the proposal.

In February 1955, the government published a White Paper on defence that made the plan public.[17] At least part of the decision was based on the idea that, if Operation Hurricane had impressed the Americans with Britain's ability to design and test a fission weapon, the next and most crucial step was to prove its ability to master hydrogen bomb technology. There was hubris aplenty during this time, according to British historian Peter Hennessy, as recounted in his lecture to the House of Lords in 2011. He said that the Chiefs of Staff argued for the H-bomb

> as a means of strengthening 'our position and influence as a world power' claiming that '[o]ur scientific skill and technological capacity to produce the hydrogen weapon puts within our grasp the ability to be on terms with the United States and Russia'. There was no way in the mid-to-late 1950s that the UK could hope to come anywhere near matching the growing nuclear arsenals of the United States and the Soviet Union.[18]

Hydrogen weapon technology, still evolving as the British decision was made, included a wide range of size, weight, yield and fissile fuels, which meant that many technical options were on the table.[19] As international moves to restrict atmospheric testing started to loom in the background, Britain strained to manage the logistics of creating such new technology quickly. The planning process began in earnest in February 1956, as part of the lead-up to Mosaic. The first major step was for Britain to start understanding the American H-bomb. Because the US had been invited to send Superfortress aircraft to Operation Totem in 1953 to observe, America reciprocated by allowing an RAF Canberra aircraft to be present at a US hydrogen weapon test at Bikini Atoll, Operation Castle, to sample the clouds from the six detonations.[20] British analysis of these samples helped the AWRE to make progress on its own H-bomb design, under the guidance of William Penney's reliable and efficient offsider, William Cook, who managed the H-bomb project. At the urging of Frederick Lindemann, Penney and Cook recruited the English-born theoretical physicist John Ward to the thermonuclear team at Aldermaston at a crucial time in 1955 for an unhappy but productive short stint.[21] Ward's American experience, particularly at Princeton, opened the possibility of unlocking the secrets of the American hydrogen bomb design that had been denied to Aldermaston by the McMahon Act.[22]

The question of where to test these powerful new weapons arose during planning, since Australia had explicitly banned H-bomb tests on its territory. Some British bureaucrats half-seriously advanced the notion of testing its thermonuclear weapons in Antarctica, partly to put pressure on two countries that were annoying them at the time over a territorial spat, Argentina and Chile. The British Foreign Office thought that strategic use of testing grounds could serve some diplomatic purposes, over and above the scientific considerations. While wanting to settle the territorial dispute in the International Court, Britain could not get Chile and Argentina to agree to go to court, so senior officials mused that it might help to increase British activity in Antarctica. The idea was outlined in July 1954 in a top-secret letter from PH Dean in the Foreign Office to Sir Frederick Brundrett, chief scientific adviser at the Ministry of Defence, later chair of the Operational Use of Atomic Weapons Committee. 'During interdepartmental discussions it was suggested that one way to alarm the Argentines and Chileans and so stimulate them to accept the International Court's jurisdiction might be to let it be known that atomic experiments or experiments with other unconventional weapons might take place in the British Antarctic sector.'[23] Acknowledging the risks, Dean said 'this would be a poor tactic unless there were real prospect of actual experiments taking place. We could not risk a bluff, which might well be called'.[24] Dean wanted to know if this was a realistic possibility, saying that if there were a real chance of the idea becoming feasible in the near future 'we should like to consider including the point in our paper for the Cabinet. We should be grateful for enlightenment on this, and for any comments you may have in this connexion'.[25] Brundrett replied,

> I think it would not be feasible to suggest the use of any Antarctic region for land explosion. I am not aware that the suggestion has ever been examined in detail but I would have expected on general grounds that such an arrangement would not be suitable for many reasons. The weather is apt to be bad in those parts.[26]

Nevertheless, William Penney did seriously consider the southern polar regions for thermonuclear tests briefly during 1954, specifically yield measurement and safety issues, and the matter was considered further by Brundrett in the Operational Use of Atomic Weapons Committee.[27]

The British government also considered trying yet again to work with the Americans at their test sites. Indeed, Admiral Lewis Strauss, chairman of the US Atomic Energy Commission, had raised this with Sir Roger Makins, British ambassador to the US, offering the Pacific site of Eniwetok. Makins kept a top-secret record of their conversation in which he noted Strauss's view that America saw as 'inevitable' the development of a British thermonuclear weapon.[28] Makins told Strauss that before Operation Hurricane the British authorities had attempted to use an American site instead of going to Monte Bello, 'but the conditions which the United States authorities had sought to impose were so restrictive that we had reluctantly decided to have our own test'.[29] Strauss claimed to be unaware of this, and also emphasised that the possible British use of the Eniwetok site was entirely his idea, and therefore not a formal offer. Strauss also suggested, according to Makins, that the US and Britain jointly develop a new Pacific testing site, to prevent 'too great a saturation of the Eniwetok area'.[30]

Britain also approached New Zealand to enquire whether it could test at the Kermadec Islands, about 1000 kilometres north-east of New Zealand's North Island. These uninhabited islands seemed suitable for what Britain had in mind. However, New Zealand Prime Minister Sidney Holland said a firm 'no' to the suggestion.[31] The Ministry of Defence reluctantly moved on to consider other possibilities, the next in line being two islands in the central Pacific: Malden Island and McKean Island (in what is now Kiribati). Moving down the list of possibilities, next on the list would be a ship test somewhere in the open ocean, 'but this would be far more difficult than a test conducted on or near to land', according to a Ministry of Defence briefing paper.[32] Before proceeding with any of these alternatives, the plan was to try Prime Minister Holland again.[33] The ministry prepared a new message 'expressing our disappointment that he has felt unable to meet our request, telling him that we shall now examine other less desirable alternatives, and hinting that if these prove unsatisfactory, we shall probably return to the charge with him later'.[34] Clearly these entreaties to the New Zealand prime minister cut no ice, and Britain proceeded with plans for Operation Grapple (1957–58) at Malden Island and Christmas Island (Kiritimati).

Meanwhile, the technical side was moving ahead under Penney's leadership. Penney and Cook had some decisions to make. Most H-bomb designs are a powerful mixture of fission and fusion technology, producing bomb blasts in the megaton range, 1000 kilotons and over. Fusion works in the opposite direction to fission, combining (fusing) light elements, a process that releases huge amounts of explosive energy.[35] These weapons generally require a fission explosion to produce the enormous pressure and heat to induce fusion. Because of the extreme heat required – around 100 million degrees – H-bombs are also known as thermonuclear weapons. British Home Office planning documents calculated that a 20-megaton blast at ground level would form a crater 1.6 kilometres in diameter and over 50 metres deep.[36] Fission-alone devices run into logistical issues when creating such huge blasts: to approach megaton range, fission bombs would have to be impracticably enormous.[37] The British fission weapons tested in Australia never exceeded one-tenth of a megaton, and most were a much smaller fraction. Despite some speculation, an actual hydrogen bomb was never tested on Australian soil. The Mosaic G2 bomb tested at Monte Bello in June 1956, the biggest nuclear device exploded in Australia, was technically a fission weapon, but one that had 'light elements as a boost' added to it, as Prime Minister Anthony Eden casually slipped into his correspondence with Robert Menzies.[38] This made it a 'boosted fission' weapon, in which the explosive power was provided by fission but was given a kick along by adding isotopes of hydrogen and lithium.[39] However, although G2 was huge by fission standards, it would be dwarfed by a genuine H-bomb. Britain tested its megaton weapons at Malden Island and Christmas Island in the Pacific, but ultimately abandoned its independent H-bomb research when it went back to working with the Americans. Its Grapple series of H-bomb tests were largely for the purpose of showing the Americans that the country had developed the means to create a fusion weapon, something it achieved, after several unsuccessful attempts, in 1958. Churchill was particularly interested in Britain mastering the technology without necessarily building its own H-bomb arsenal – the art, not the article, as he put it.[40]

Just one month after Operation Hurricane, on 1 November 1952, the US successfully tested its first hydrogen weapon, codenamed Ivy Mike, at Eniwetok Atoll. This device was extremely unwieldy, and the test was (like Britain's Hurricane bomb) more for show than for establishing the design parameters of an actual weapon.[41] Just as Britain was starting to make some headway in atomic weapons technology, America (and, soon after, the USSR) was sprinting off again beyond the horizon. Churchill took a keen interest in Ivy Mike.[42] Britain, as ever, had one eye on proving to its ally that it could keep up, and indeed one US Congressman had witheringly described what Britain could bring to any nuclear negotiation as trading 'a horse for a rabbit',[43] a humiliating insult to a once-dominant colonial power. However, as official UK historian Margaret Gowing relates, initial scepticism in the US did give way to admiration about what Britain had been able to do in straitened circumstances. The British gamble, fuelled by a certain amount of hubris, appeared to be paying off, if American admiration could be taken as currency.

As 1953's Operation Totem showed, the UK had been struggling to produce the raw materials for a modest 'conventional' fission bomb arsenal, so how could it manage the technological requirements of the H-bomb? As it happens, this form of weaponry can be relatively less expensive than fission bombs. Historian Kevin Ruane said that the government's Working Party on the Operational Use of Atomic Weapons 'had already assessed that the country could produce annually around ten hydrogen bombs of the 5–10 megaton class, starting in 1959'.[44] The particular design adopted for some of the Grapple tests by the British bomb makers, in which the heat generated by a fission component would cause a fusion reaction, 'was regarded as essential for producing the necessary number of weapons from the limited amount of fissile material likely to be available'.[45] The smaller yield requirement for the fission component meant that severely limited and expensive supplies of plutonium could go further in this style of weapon. In 1956, weapons-grade plutonium cost £40 000 per kilogram.[46]

The relative cheapness of the H-bomb, and its (potential) simplicity, swung the economics of British nuclear weaponry over to a smaller number of more powerful weapons. Cockcroft at Harwell and Penney at AWRE were on board, and they prepared a brief in March 1954 for the government as ministers were about to decide the future of the bomb program. Kevin Ruane wrote that their 'brief began with two arresting sentences: "Thermonuclear weapons are undoubtedly simpler to make than scientists thought. Great skill may be required, and precise engineering, but a vast industrial effort beyond that required for producing plutonium and [uranium-235] is not needed"'.[47] So another line of demarcation was drawn, and the UK nuclear program took a new path from 1954. Hurricane at Monte Bello and Totem at Emu Field were quickly consigned to the old technology era as Britain gambled its future defence capabilities on the untold energy released when isotopes of hydrogen and lithium are fused.

Churchill for his part advocated for the nuclear superiority of the NATO powers, with a diverse new military arsenal that was not weighed down just with the hideous might of the H-bomb but made use of all the new technology. He saw fission weapons as part of conventional armoury, and in effect disguised within it:

> Smaller varieties [of atomic weapon] should be designed capable of being carried by bombers from any large-scale airfield without the necessity of making the tell-tale prolongation of the take-off concrete track. Bases should be established in as many parts of the world as possible. Camouflage should be used to convey the impression of numbers as well as to disguise reality.[48]

Post-1954, as Britain pursued its thermonuclear deterrent, the country also entered a period of contradictions. Its technology for both fission and fusion weapons was still in its infancy and required testing, yet the country was also actively pursuing international mechanisms to ban atmospheric nuclear testing. These efforts, combined with those of the US and USSR, resulted first in the three-year moratorium on atmospheric testing enacted in Geneva in 1958, and later the partial test ban treaty that came into effect in 1963. While working doggedly behind the scenes to design and build a nuclear and thermonuclear arsenal, the British were also at the forefront of actions to stop these weapons being tested in the atmosphere. According to historian John Walker, '[p]olitically the UK faced what appeared to be irreconcilable objectives: the need for thermonuclear weapons and the need to be seen to address domestic and international public concerns over the health, environmental and international security effects of nuclear tests'.[49]

By early 1954, Winston Churchill – who had proudly, boastfully overseen the early tests of Britain's A-bomb – was failing. According to scholar Henry Pelling, quoting Churchill's personal secretary Jane Portal, he kept going despite significant infirmity because 'he thinks he has a mission on three subjects – Russia, Egypt, and the atomic bomb'.[50] The rise of Russia as the West's Cold War enemy had many years to play out, long after Churchill died in 1965, while the Suez situation in Egypt would be for his successor, Anthony Eden, to take on, with unfortunate consequences. Churchill would live to see the third of these ambitions realised, as Britain spent the 1950s developing its own nuclear arsenal, only to abandon its own designs and adopt those of the United States by decade's end. The 80-year-old statesman made the bomb the theme of his final speech to the House of Commons, on 1 March 1955, fittingly since nuclear weapons development had dominated his second term as prime minister (and, in a way, his first as well when he pushed for Britain to be involved in the Manhattan Project, having overseen its research and development forerunner, known as Tube Alloys). Not only did he outline the current state of armament among the three nuclear-armed nations, he also urged disarmament, for fear that the weapons his country had helped develop might be used again. Churchill expressed a view that Britain had to maintain a system of deterrence, and this became the dominant narrative as the country moved to thermonuclear weaponry. Churchill used his final speech to set out his philosophy:

> Unless a trustworthy and universal agreement upon disarmament, conventional and nuclear alike, can be reached and an effective system of inspection is established and is actually working, there is only one sane policy for the free world in the next few years. That is what we call defence through deterrents.[51]

The notion of deterrence, what later morphed into 'MAD' (mutually assured destruction) as the world's nuclear stockpile grew and grew, was essentially the rationale for Operation Mosaic, given its role in the British H-bomb program. In his speech, Churchill intoned the memorable line: 'We have entered a period of transition in which the past and the future will overlap'.[52] At that moment, as Britain was poised to develop its own doomsday weapon, its head of government was about to undergo transition too, with Churchill departing the role on 6 April 1955 and the erstwhile foreign secretary and deputy prime minister, the dapper 57-year-old Anthony Eden, stepping in. Eden would lead the Conservative Party to an election win only one month after becoming prime minister.

Eden had many crosses to bear during his brief tenure, between 1955 and 1957. His Waterloo was the Suez crisis, which occurred in the same year as Mosaic and which shortened his prime ministership, aggravated by his poor health. (Eden ordered a military response when Egypt nationalised the Suez Canal, a foolhardy venture that backfired spectacularly and angered the Americans.) However, Eden played an active political role in the advancement of British nuclear weapons and was a key figure in the development of the H-bomb – and in moves to stop further H-bomb development. Eden did not have the same vivid interest in atomic weaponry as his predecessor, although his years as foreign secretary reinforced his belief in the need for Britain to maintain 'great power' status. Also, by 1955 the British government's most enthusiastic nuclear barracker, Frederick Lindemann, was out of the picture (he retired as paymaster-general in 1953 and died in 1957). Indeed, Eden's successor, Harold Macmillan, was more vocal during his premiership about the benefits of nuclear weaponry and energy. But history placed Anthony Eden into the picture just when Britain was pushing forward with its H-bomb program, and he therefore had a role in the events that played out at those faraway isles.

Upon Eden's ascension to power, global concerns about the upsurge in testing of both fission and fusion weapons was growing. The word 'fallout' was often in the news, as the risks of atmospheric testing became familiar to the general public. Eden inherited the British H-bomb program championed by Churchill, but he would have to politically and morally justify the inevitable harm that these high-yield weapons tests would cause. This was not the only uncomfortable position Eden was to find himself in during his 20-month tenure, as the Suez crisis tore like a tornado through his government. Suez further dented Britain's strained relationship with the United States, with President Eisenhower refusing to support Britain's military actions against Egypt.[53]

Anthony Eden would later reaffirm the British government's view that the decision to pursue the H-bomb was meant to keep the world safer when he said that he and the government 'regard our manufacture and possession of the hydrogen bomb as a deterrent to war'.[54] He added that Britain was 'not prepared to accept agreements which would put the United Kingdom in a position of decisive inferiority to other great Powers, a position which is not justified by the state of our scientific knowledge and resources'.[55] However, he flagged willingness to discuss with the US and USSR ways to slow the arms race and reduce the growing harm caused by atmospheric testing, something already called for by India and the USSR, and which would soon be championed by President Eisenhower. These measures would come into force in October 1958 with the moratorium agreed in Geneva by Britain, the USSR and the US, all three governments having first proved their ability to create a megaton weapon, and not before. These developments were taking place against a backdrop of rising public uneasiness about nuclear testing, and recognition that the days of unfettered atmospheric testing were numbered. Project Sunshine, which began in 1953 as a classified survey by the Rand Corporation, amended its findings in 1956 and released a declassified version of its report in 1959 about the rising risk of radioactivity released by testing. The message was clear – atmospheric testing was placing large quantities of the bone-seeking radioactive element, strontium-90, into the atmosphere, and it was falling to earth where it was readily taken up in the biosphere.[56] For each nominal (20-kiloton) atomic weapon, 20 grams of strontium-90 was released.[57] As the report noted, that was 'a considerable quantity'.[58] By 1953, a total of 19 nuclear weapons had been tested by the three nuclear powers, but many more were planned in 1954. (Project Sunshine became controversial when it was made public that it had analysed tens of thousands of bone samples, particularly from dead babies and infants, without the consent of the families.) As public heat intensified, it became politically advantageous for Britain to openly advocate for a tripartite test ban agreement.[59] William Penney recalled decades later that 'warnings kept on coming from the Foreign Office that a test moratorium had ... begun to look possible'.[60] As the road to Geneva became clear, the British timetable was set. The country had to get its megaton weapon design successfully tested before the shutters came down.

One early task for Eden was to make contact with Robert Menzies to secure the Monte Bellos for Mosaic, once the AWRE had determined that the operation had to be held there. The negotiations were potentially more delicate than usual, given growing political sensitivities in Australia. On 16 May 1955, he told Menzies that 'certain experiments' were required in early 1956, ahead of the thermonuclear tests at an as-yet unconfirmed South Pacific location in 1957.[61] The oft-quoted line that gave a hint at what Mosaic would involve was used by Eden in this correspondence: that the experiments would involve atomic explosions 'with the inclusion of light elements as a boost. It would of course be made clear in any public announcement that the explosions were atomic and not thermonuclear'.[62] The emphasis on the letter of the agreement with Australia was clearly intentional, and it is possible that Eden himself did not possess enough technical knowledge to understand the subtle differences between weapon designs. He also made a clear pledge that neither of the Mosaic bombs would be any more than two and a half times the 25-kiloton explosive yield of Hurricane, meaning that they would not exceed 62.5 kilotons, a political undertaking that may have influenced later understandings of the yield of Mosaic G2. He further asserted that the explosions would be conducted on towers 'to reduce contamination and fall-out would be less that one fifth of that of the Hurricane bomb'.[63] The Mosaic program was highly experimental and its details were in a state of flux throughout the planning phase, including the number of bombs that would be detonated. However, its objectives and tight timeframe were firmly linked to later plans for the South Pacific British H-bomb tests.

While agreeing in principle to the proposal, Menzies indicated in his response the following month that he would need to know more about the commitment expected of Australia. On the basis of the information in Eden's message, Menzies said that it was not possible to accurately assess the extent to which Australian resources could be involved in logistical support for the trials.

> [In] determining our ability to assist, we must take note of the cumulative effect of our existing and prospective commitments such

> as the provision of part of the strategic reserve in Malaya and our joint operations at both Woomera and Maralinga, and view these in the light of the resources of the Australian Services and the current shortages of manpower.[64]

Menzies suggested that Eden send representatives to Australia to discuss in detail the requirements, including the finances. He wrote:

> I notice from your description of the nature of the shots to be fired, that each would give a yield considerably greater than in the case of the bomb fired in the 'Hurricane' operation. In view of this, I ask that the most meticulous care be exercised in the scientific checking of safety measures to ensure the safety of people and animals on the mainland.[65]

He concluded that while he was anxious to help out, it was only fair to alert Eden that Australia might not be able to provide everything the British were hoping for.

Eden wrote back soothingly in early July, assuring Menzies that Mosaic would cost a tenth of what Hurricane had cost, saying 'our needs will be relatively modest'.[66] He advised that Admiral Hugh Martell, commodore of the Mosaic task force, would lead a British delegation, departing the UK on 18 July and arriving in Australia on 22 July, to sort out the details and answer any questions from the Australians. Charles Adams, who was appointed by William Penney to be the scientific director for Mosaic, and Paddy Menaul, who would run Mosaic air operations, were included.[67] In response to Menzies' concerns about safety measures, Eden said, 'I assure you that safety measures will be most meticulously carried out and that as before we should welcome discussions on safety checks with your scientists. We shall be glad to have some senior Australian Service officers as observers'.[68]

Eden developed a close relationship with Robert Menzies. They maintained a warm correspondence even after Eden was forced into retirement. Eden planned a trip to Australia, as well as New Zealand, in January–February 1957 to further cement the cordial relationship he had with the Australian leader. Indeed, he announced this trip in the House of

Commons the day after Mosaic G1 was detonated, 16 May 1956.[69] However, in December 1956, under pressure from his own party, the trip had to be cancelled. The Suez crisis was threatening to engulf his premiership, and he needed to stay home. The secretary of state for Commonwealth affairs, Alec Douglas-Home, sent a diplomatic but firm personal message to him on 4 December 1956:

> It seems clear that the next few months are bound to be a period of very great difficulty, not merely for the party, but from a national point of view. We think that it would be wrong for you as Prime Minister to be out of the country for the greater part of January and February.[70]

Douglas-Home suggested 'softening the blow' by letting the leaders of Australia and New Zealand know immediately.[71] The following week, Douglas-Home wrote to Menzies, saying, 'In view of the general situation, both on the international front and here at home, where the country will be faced with many serious problems over the next few months, the Prime Minister has decided that it will not after all be possible for him to keep to his plans'.[72] Had he made the trip, Anthony Eden would have been the first British prime minister to have visited Australia while in office.

The cost of the Mosaic program was a subject of questions in the UK parliament. Just over a week before G2, Labour member and prominent pacifist Emrys Hughes asked the minister for supply, Reginald Maudling, what the cost of the program was. When Maudling refused to answer, Hughes wondered whether he was 'trying to hide the facts from the British taxpayers more than from any potential enemy?'[73] Maudling replied, 'I am afraid it is the best I can do. I have looked at it very carefully. There is no doubt that if one gives figures of this kind they do enable other countries to assess the size of our nuclear effort'.[74] Then, at the moment when G2 was being tested, Eden was on his feet in Commons, letting it be known that moves were afoot to abolish all nuclear testing and that a Disarmament Commission would meet soon at the United Nations to advance the idea. Labour member Stephen Swingler protested the 'colossal wastage and potential dangers of further tests'.[75] Eden responded that the Disarmament

Commission was the place for such discussions to take place.[76] The tests had a momentum that couldn't be stopped, set by the coming moratorium. Mosaic would have to go ahead before Grapple, so that Grapple could prove to the Americans that the British were in the H-bomb race. So go ahead it did.

CHAPTER 6
AUSTRALIANS AT MONTE BELLO

> … the United Kingdom can be told to stuff their bomb up their jumpers.[1]
>
> ATTRIBUTED TO THE AUSTRALIAN DEFENCE COMMITTEE, SEPTEMBER 1952

> Having committed us to the deadly atomic tests off our west coast, and having condoned USA's Pacific H-test, Menzies sailed away on a leisurely voyage to Britain. He was thousands of miles away when the Monte Bello bombs went off.[2]
>
> SYDNEY *TRIBUNE*, 27 JUNE 1956

Australia had little role at Operation Hurricane, and not much more at Mosaic. The involvement of the Australian government under Robert Menzies, and the main Australian-associated scientists, was often actively opposed by some elements in the British establishment, and always kept to a bare minimum. Near-total exclusion was the British *modus operandi* throughout its atomic tests in Australia, and the emphasis in communication between the nations was on reassurance rather than transparency. For a while, the Australian government went along with it. There was no doubting the propaganda benefits of the Menzies government saying yes to the Brits for atomic testing, particularly in the early days. After all, the great anglophile prime minister was allowing Australia to do its bit for future Western security. The immense secrecy around the tests did still leave room for the release of vetted information designed to put both governments in a shining light and effectively hide the dangers from the Australian people. In this, Operation Hurricane was completely successful, Operation Mosaic less so.

Plenty of self-interest was involved. Both Operation Hurricane and Operation Totem at Emu Field were deemed by the Australian government to have publicity (and, therefore, political) dividends worth the pain of having to co-operate with a stubbornly secretive and imperious British government. These early British atomic tests in Australia, held so soon after the end of the war, were moments of pride and celebration in Australia. The nation was participating in a great postwar security enterprise that would ward off the growing threat of the West's new enemy: the Soviet Union. Patriotic sentiments rang through the media coverage at the time. Operation Hurricane took place in an atmosphere of nationalistic zeal, with few dissenting voices raised. Likewise Operation Totem, held in the centre of Australia, despite the unfathomable suffering it caused the Aboriginal people of the area, was nevertheless an occasion of national celebration. Operation Mosaic was a different story. By 1956, the patriotic fever had broken, and Australia was increasingly nervous about what the Brits were up to. Mosaic G2 accelerated the process of public concern, with its torrent of stories about fallout crossing the mainland. That year saw a decisive shift in public sentiment as the British test program was pivoting to hydrogen weaponry, and as Maralinga was being readied as the 'permanent' test site at which the fission components of the H-bomb would be among the devices tested.

The Australian media were mesmerised by the early atomic tests. A typical newspaper response to Operation Hurricane can be seen in an editorial from the *West Australian* of 4 October 1952 – the day after the Hurricane shot at Monte Bello – to accompany the extraordinary reporting of Norm Milne and others. It stated:

> The real significance of the Monte Bello explosion lies at this moment ... in the simple fact that it occurred. It gives the world the indisputable proof that Britain has the material, the skill and the installations for the independent production of atomic weapons and that she will yield the initiative to none.[3]

This lengthy editorial continued in the same vein, talking about 'profound satisfaction' and 'a strong sense of pride', reflecting the tone of much

Hurricane coverage. It concluded: 'the Monte Bello explosion reverberates with a vastly increased assurance of British Commonwealth power and defensive security'.[4]

As the 1950s wore on, though, public sentiment soured. Key Australian players such as the respected Australian defence scientist Professor Leslie Martin did not hide their disdain behind the scenes, while keeping up patriotic appearances in public. Martin in particular was treated shabbily by the British, used as a prop to give the suggestion of official Australian oversight while not actually allowing any Australian too close to the centre of the tests. Martin's umbrage at being denied crucial information necessitated some fancy footwork by British officials to get him back on side. The Royal Commission later found 'some reluctance on the part of the British authorities to allow any Australian presence'.[5]

Some concerns about the painfully obvious lack of Australian participation were voiced, quietly and without much follow-up. Labor member William Bourke asked a question of Menzies' defence minister, Philip McBride, in the House of Representatives on 29 August 1952, five weeks before the Hurricane test. He requested that the minister indicate the extent of Australia's participation in the Monte Bello test. 'Is it correct, as has been reported, that the British authorities who are undertaking the tests are reluctant to allow active Australian participation in their work? If that be so, what is the reason for their attitude?'[6] McBride responded by indicating that the conduct of the tests was entirely in the hands of the British government.

> If the British Government should need further help, the Government would comply with any request it made. However, the British Government is the authority responsible for the tests, and will itself decide the nature of the tests to be carried out, who shall attend them and the information that will be released as a result of them.[7]

A delicate line had to be negotiated here, between acknowledging that Britain was in charge of the tests but ensuring that the country hosting the tests had meaningful oversight.

William Penney sought views on the involvement of the Australians from a reluctant Frederick Lindemann, in a secret note on 27 February 1952. The issues involved were both scientific and political, and he wanted Lindemann to give the political lead.

> Clearly we could not possibly allow any Australian to see the weapon, or the weapon functioning equipment or the radio-chemistry. On the other hand I see no risk in allowing two or three scientists, specially nominated and guaranteed by the Australian Government, to see
> (a) Concrete shelters, before and after the explosion.
> (b) Aircraft components ...
> (c) All photographic equipment, and some censored photographs and films.
> (d) All blast measuring equipment, and some censored results.
> (e) Contamination surveys, after the explosion.[8]

Penney said that 'there must clearly be a difference in our behaviour towards (1) Canadians, (2) Australians, (3) other dominions'.[9] The Canadians were trusted long-term atomic allies, but the Australians were entirely inexperienced in this area. The British stance was emphasised by EL Sykes from the CRO, in a secret note to Frank How at Supply: 'The work to be undertaken by the Australian scientists is of very low security classification. These people will not be close enough to the site during the actual trial to be able to draw any scientific conclusions from it. (Will they even see the "puff of smoke"?!)'[10] The British did not want Australian involvement and did not need it. However, politics dictated that allowances had to be made for the sake of appearances.

Behind closed doors and somewhat belatedly, the Australian government had started dropping hints to the British that some direct Australian involvement needed to be arranged. A meeting of the Australian Defence Committee on 28 August 1952, chaired by the secretary of the Department of Defence, Sir Frederick Shedden, decided that Leslie Martin would be the preferred choice to be offered to the British. The committee specifically required that he be present at Monte Bello so that 'he might be fully acquainted with the details of the tests'.[11] As pressure mounted for a token of Australian involvement, British thoughts had also turned to Martin, a well-regarded University of Melbourne physicist who had strong links with the defence community. He had been defence scientific adviser since 1948 and chairman of the Defence Research and Development Policy Committee, and would become one of the founders of the Australian Academy of Science in 1954. He had also served on the Commonwealth Advisory Committee on Defence Science at the same time as William Penney. George Davey at the British High Commission in Canberra had contacted the assistant secretary in the Prime Minister's Department, Allan McKnight, to indicate British interest in Martin.[12] Although the British, for some reason, had assumed that Martin had signed the *Official Secrets Act*, he was an Australian working in Australia. In the minutes of the 4 September 1952 Defence Committee meeting, this was clarified:

> In regard to the assumption by the United Kingdom authorities 'that Professor Martin is appropriated covered by the Official Secrets Act, etc', it is pointed out that he is the Defence Scientific Adviser and comes within the provisions of the Crimes Act which corresponds, in part, to the British Official Secrets Act.[13]

Penney was comfortable with the choice of Martin. He wrote to Lindemann on 15 August 1952, describing Martin as the best man for the job, having already demonstrated 'an excellent spirit of collaboration and a realisation of the security aspects of the matter'.[14] Lindemann was against any Australians being involved, although he ended up relenting. He tried to block Alan Butement from Supply. A letter from a member of his office staff, EJS Clarke, said that the paymaster-general had realised

> that it is not possible to refuse to accept Butement unless lack of accommodation provides a valid excuse, but he is not too happy about this addition to the team. He would, therefore, like a personal cable sent to Penney saying that while we will agree to this proposal we assume that he will not have access to vital efficiency data.[15]

Including Australians was looking like being necessary to maintain Australian co-operation. A letter from Frank How in the Ministry of Supply to Robert Saner in the Foreign Office noted that after the Hurricane test at Monte Bello, the tests were set to move to central Australia.

> The Commonwealth Government are likely to be nervous about allowing the use of a site in the heart of the continent for atomic weapons tests, and may have to face criticism from their own people. It is obviously desirable that one of their own scientists should be able to advise them from first-hand knowledge, and it seems right to use the Monte Bello test as an opportunity for indoctrinating such a scientist.[16]

In those days, the word 'indoctrination' did not have the more sinister connotation it now has; it referred more to instructing.

The Ministry of Supply official sounded a note of caution, however, saying that the 'best way to prevent the Australians from feeling that we are "attempting to use their land but at the same time keeping them out" would be to ask them to make a nomination for this purpose, but we cannot risk their unfettered choice'.[17] The British saw Martin as a safe option. Neil Pritchard from the CRO followed up this correspondence with his own letter to Saner, endorsing the idea of allowing Australia to have Martin attached to the project. 'As Professor Martin is Scientific Adviser to the Australian Department of Defence, I am sure that we must approach the Australian Government, though we can leave it to our people on the spot to decide whether to do so before Martin has been sounded or afterwards'.[18] Pritchard indicated that they couldn't make a move, though, until Paymaster-General Frederick Lindemann, Churchill's closest confidant, had given his approval. Lindemann agreed, if reluctantly.

A member of the UK Services Liaison Staff (UKSLS) subsequently visited Martin in Melbourne to ask him about attending Hurricane as an official observer. William Penney also sent a personal message to him, alerting him confidentially and wanting him to know about the invitation before the Australian government was approached. Penney said that ministerial approval

> has now been obtained here for extending to you a cordial invitation to go to Montebello as a member of the Health Physics team. You will be given access to all weapons effects, all measuring apparatus, and all details of the layout of the site. In particular we should like you to work with the meteorologists following movement of contaminated clouds.[19]

The grab-bag list of what he would have access to had carefully avoided mentioning what he would *not* have access to, something that would quickly become a sore point. Penney asked that, should he accept this invitation, he travel from Melbourne with Ernest Titterton to Monte Bello. Titterton, a Manhattan Project veteran, was the newly appointed head of nuclear physics at the Australian National University. At this point the test was only weeks away; it was painfully obvious the invitation was an afterthought. Martin had university commitments that might make getting away difficult.

Initially, Martin was 'delighted and tentatively accepted'.[20] Early enthusiasm turned to anger, in the account of an UKSLS official, Captain Christopher Hutchinson, who briefed the UK deputy high commissioner in Australia, Ben Cockram, about interactions with Martin. The invitation was discussed at a meeting of the Australian Defence Committee on 4 September 1952, with no British representative present. By then, Martin's invitation to play a carefully managed role in Hurricane had been referred to the committee by the Prime Minister's Department. The invitation contained details of 'certain restrictions' on what Martin would be allowed to do. Hutchinson told Cockram that there was 'a most unhappy scene because of the restrictions, the Committee deciding to recommend to the Ministers that the invitation as it stood was an insult and should be declined. One of the terms was apparently "the United Kingdom can be told to stuff their bomb up their jumpers"'.[21]

The minutes of the Defence Committee meeting do not specifically record that colourful quote, but Australian displeasure comes through the bureaucratic language all the same. They do indicate that the 'limited participation' offered by the British was not acceptable.[22] The British invitation had specified that Martin would not have access to any information about the Blue Danube weapon design or any measurements arising from the test. This apparent tokenism angered Martin and the rest of the committee, for three main reasons: the significant input of Australian resources and use of Australian territory at the Monte Bellos; the importance of the test program to defence research; and the need for Martin, as defence scientific adviser, to have as much access as possible to information so that he could advise the Australian government, especially as Hurricane was likely to be the first of many tests.[23]

Martin was treated differently to Titterton and Butement in the hierarchical Hurricane arrangements: Titterton was already known to the nuclear test authorities as one of their own and was privy to significantly more insider information, while the New Zealander Butement had worked in the British civil service, including in the Ministry of Supply, before joining its Australian counterpart (although he would not enjoy the same access as Titterton, and was there under sufferance from Lindemann). Martin's credentials, on the other hand, were entirely Australian. According to scholar Tim Sherratt, this made him 'a crucial observer to the Australians, and an uncomfortable necessity to the British, who sought to protect their own interests by imposing limits on his involvement'.[24] Martin may have maintained some lingering resentment after Hurricane since he refused to allow his research students to be involved in the Emu Field tests in 1953, prompting Titterton to include his own students instead. He did, however, agree to being appointed as inaugural head of the Australian Atomic Weapons Test Safety Committee in 1955, in time for Mosaic. None of the Australian observers at Hurricane were aboard the operation's flagship, HMS *Campania*, with Penney. Instead, they were accommodated on HMS *Narvik*.

Concerned at losing Australian goodwill, UKSLS asked Sir John Cockcroft to write a placating letter to Professor Martin. Cockcroft, who was deeply involved in his country's atomic energy enterprise through his role as head of the AERE, happened to be visiting Australia at the time. Sir John's letter arrived at the UKSLS on 9 September and Hutchinson hand-delivered it to Martin, 'feigning that I was unaware of the author or contents ... I also, of course, feigned ignorance of any change of Martin's attitude since he had previously tentatively accepted the invitation through me'.[25] Martin recognised Cockcroft's handwriting, but was unable to decipher it. 'This was fortunate in that he called for my assistance. I read the letter right through to him and, believe me, it was a masterpiece of persuasion though in no way giving away the fact that United Kingdom representatives were aware of the previous Defence Committee scene.'[26] No doubt flattered by attention from such an eminent physicist, Martin wavered. Hutchinson took the opportunity to reel him in: 'I pointed out that even if feelings had been hurt the refusal would not only deprive the United Kingdom of his assistance and Australia of the experience, but would also make invalid any further Australian criticism of lack of opportunity to be scientifically associated with the test'.[27] That did it; Martin agreed to be involved and started making arrangements to attend the test the following month, including deciding what to wear. He accepted the restrictions placed upon him, and agreed to a nebulous role as an 'observer'. In a handwritten note to Penney that same day (also hard to decipher), Cockcroft conceded that the Australians were 'a little sore about the way co-operation in Hurricane has been carried out'.[28]

In parallel with the wooing of Leslie Martin, head of the Prime Minister's Department Allen Brown had made contact with Ernest Titterton at the ANU. Titterton gave testimony later at the Royal Commission that he had immediately confided in prominent Australian physicist Mark Oliphant, who had given his blessing.[29] Oliphant had been Titterton's supervisor at the University of Birmingham, and they were close – indeed, Titterton referred to his love of Oliphant in his testimony.[30] Despite their closeness, though, they did not work together on the British tests in Australia. Oliphant was cast aside due to questions posed by the Americans about his security soundness. Titterton presented no such issues, however. He was known to the British nuclear establishment, and had been part of the Manhattan Project, initiating the world's first ever atomic weapon in the Trinity test in New Mexico. He had also worked at the AERE at Harwell.[31] He claimed, though, not to have known anything about being involved with the British tests in Australia until he was approached by Brown.[32] This is unlikely, given his strong ties to the British nuclear establishment. Titterton reported that, after being appointed as an observer, he frequently hopped on his bicycle to ride over to Parliament House from the ANU to chat with Robert Menzies about the forthcoming Hurricane trial.[33] During these discussions,

Titterton discerned that Menzies was party to some information about the test, but clearly did not understand it fully. 'He said to me – on one or two occasions – what does so and so mean, and the moment he said it, I knew that he had been informed of something in the UK which, having very little scientific training, he did not understand.'[34] From the start, Titterton was better informed about British plans than the other Australian observers, and had regular discussions with William Penney about technical matters, including expected yield.[35]

Titterton and Butement joined Martin as Operation Hurricane observers, a role with no specific duties or responsibilities. The acting British high commissioner wanted no fanfare: 'in view of the fact that Professor Martin's association with the test was being kept secret, it would be a good idea if no press announcement or publicity whatsoever were given to Mr. Butement's attendance'.[36] Titterton's role was, however, known publicly. As Martin had feared, the 'Australian scientists were ... given a public and political rather than a scientific role'.[37] All three would continue in the same capacity at Operation Totem at Emu Field in 1953, again having an unclear brief, and largely used by the British as evidence of Australian involvement, with only Titterton doing work of any substance (he and his students carried out a program of experiments at Emu Field). As the investigative journalist Brian Toohey wrote, 'Australian scientists and public servants couldn't give informed advice on the wisdom of agreeing to the tests because none were briefed by the British'.[38] In testimony to the Royal Commission, William Penney said that Martin, Butement and Titterton could have had plenty of power at Hurricane, including veto power, and indeed could have done their own yield measurements if they had wanted to, despite evidence that the Brits wanted the Australians (especially Martin and Butement) to know only the bare minimum. Penney said that if 'Professor Martin and Mr Butement felt that I had not told them the yield and they certainly wanted to know what it was, it would not have been difficult for them to measure it for themselves, because we were certainly willing to show them blast measurements'.[39] When pressed by counsel assisting the Royal Commission Peter McClellan about why Martin in particular had not been involved in the decision about when to fire the weapon, Penney wondered aloud whether Martin had actually said anything to him about it, implying that Martin had not been upfront about feeling excluded. McClellan countered that Martin's frustration led to the creation a few years later of the Australian Atomic Weapons Tests Safety Committee (AWTSC). Penney did not disagree.[40]

By the time of Mosaic, the Australian government was starting to get a bit more demanding. The new AWTSC had made negotiations between the two governments somewhat less easygoing. Now, for the first time, the Australian government had a formal mechanism for involvement in the British tests, one that came with an explicit veto on firing. This veto was never exercised to its full extent, although its existence was a source of some tension between commodore of the Mosaic task force, Hugh Martell, and the Australian government, as well as members of the Safety Committee. The contamination and security mess left behind after Hurricane also gave the Australian government pause. A secret memo from Captain FB Lloyd to William Penney alluded to this in 1955. Lloyd mentioned the withdrawal of HMAS *Hawkesbury* from security patrols of the islands, and the consternation this caused in Canberra, saying, 'the Australian Government were very concerned at this situation and ... it was having an influence on their reaction to our present proposal for further trials'.[41] At this point, the British establishment harboured deep suspicions about Australian involvement. In July 1955, Gurth Kimber from the UK High Commission in Canberra gave voice to these in a secret memo to Lloyd, in which he decried moves by Howard Beale to have joint control of the new Maralinga site as 'unreasonable'.[42] He also labelled Butement as engaging in 'empire building', adding 'there are therefore the makings of future difficulties in the present situation'.[43]

Complicating the more difficult relationship, Mosaic was to have features associated with the development of the hydrogen bomb. William Cook wrote to Sir Richard Powell at the Ministry of Defence in April 1955 on this matter, saying, '[w]e cannot avoid telling the Australians that the bombs fired will contain small quantities of thermonuclear material'.[44] The cover story for this experiment, which Cook suggested in an attached draft message could be conveyed to the Australian government, was to boost the yield of small fission weapons, a plausible goal. However, as Cook continued, '[t]he correct reason is to obtain early information of the likelihood of success of the 1 megaton weapon as conceived at present'.[45] Cook gave Powell the option of linking Mosaic explicitly to the hydrogen bomb program, or just continuing with the vague information about boosting smaller fission weapons. The British government chose the latter approach.

At the fourth meeting of the AWTSC held in Melbourne on 10 January 1956, the committee had safety arrangements for Mosaic on its agenda. Commodore Martell had been approached earlier for his input, and his reply indicated that 'the degree of contamination would be comparable with that of "Hurricane" and that the procedures similar to that followed after "Hurricane" trials could be used namely, placarding of danger areas in several languages, and full Press publicity of the dangers involved'.[46] The meeting agreed that the highly contaminated areas should be fenced off with cyclone netting and 'amply placarded'.[47] Leonard Dwyer from the Commonwealth Bureau of Meteorology (and later member of the AWTSC) provided meteorological data from the previous three years. These showed that, on average, only two occasions per year had weather conditions suitable for atomic testing.[48] The meeting minutes note 'how very important it is to have the correct basis for defining the meteorological conditions and safety sectors'.[49] The meteorological issues would become something of a sore point between the British and the Australians at Mosaic.

Commander Martell required members of the AWTSC, as well as Leonard Dwyer who was joining the group, to assemble in Perth on 10 May 1956, following their TAA flight from Essendon airport, Melbourne.[50] After overnighting in Perth, they flew out of RAAF Base Pearce on an air force plane to Onslow. Advice from the AWTSC secretary on what to wear specified that '[k]haki tropical kit will be suitable for day wear and dark trousers, white shirt and a tie will be in order as mess dress'.[51] The AWTSC contingent included Cecil Eddy, director of the Commonwealth X-Ray and Radium Laboratory, who contracted pneumonia during this trip and died in Perth on 27 June 1956, aged 56. The minutes of the eighth AWTSC meeting acknowledged that his involvement in Mosaic led to the tragedy, saying that his early death resulted from 'his conscientiously attending to his duties as a member of the Committee at a crucial time in Operation Mosaic at the expense of his health'.[52]

Ernest Titterton, member of the AWTSC for the first two years of its existence, then chair until it was disbanded in 1973, was a reliable cheerleader for the tests. He featured frequently in the media, making assurances about safety and providing positive background about the tests, while Butement and Martin were kept out of the public eye. The day that Mosaic G1 was detonated, he was to be found in the *Sydney Morning Herald*, in a question-and-answer format article, explaining why Britain had to test in Australia. He smoothly ran the British line about the (alleged) unsuitability of the British Isles.

> Because of the hazards from radio-activity which follows atomic weapons explosions the tests are best carried out in isolated regions – usually a desert area or in the centre of a little-used ocean ... It would ... be hazardous to explode even the smallest weapons in the U.K., and it was natural for the mother country to seek test sites elsewhere in the Commonwealth.[53]

Without hinting at the real reason the British were returning to Monte Bello in 1956, he made the point that Monte Bello and the soon-to-be-commissioned Maralinga were 'complementary', each offering technical and logistical benefits not available at the other.

Behind the scenes, Leslie Martin and his Safety Committee colleagues did not always manifest the same united front with the British authorities. In particular, the requirements of the AWTSC regarding meteorology were a source of some dissension. As always, the British wanted to control everything. However, the Australians felt more keenly the requirement to keep their own country safe. The points of difference were severe enough to require a high-level meeting, scheduled during the interval between G1 and G2. Charles Adams and Ieuan Maddock, along with the British meteorologist Albert Matthewman, travelled to Melbourne to meet with Leonard Dwyer and Henry Phillpot. Phillpot worked with Bob Southern and Dick Fotheringham aboard *Narvik* as leader of the meteorological team. The main item for discussion was the all-important meteorological data, which could make or break a firing time. According to Arnold and Smith, the 'British thought the Australian method of prediction unsatisfactory and unrealistic; since the meteorological conditions they were seeking were elusive, excessive constraints could make them unattainable. Dwyer made it clear that, for his part, this consideration was secondary to maintaining absolute safety'.[54]

The Australians were more in tune with the growing political issues associated with the tests. Straight after G1, Martell sent a handwritten note to Penney, outlining some of the issues that came through from a meeting he had had with the Safety Committee. 'From their attitude ... it was clear that they are very perturbed about the Australian political situation ... Both Martin and Titterton show this apprehension most markedly.'[55] The Safety Committee pressed Martell for more details about the forthcoming G2 test, and he reiterated the general information supplied to Menzies, that the inclusion of light elements would provide valuable scientific data they needed as they planned Operation Grapple, and that they would be supplied with projected fallout patterns based on the inclusion of thermonuclear elements in the bomb design. Martell said, 'I ought, I think, to have stone-walled on the lines that such information as US was prepared to give had already been given to Mr Menzies and that they should get their information from him'.[56] However, given that the dreaded AWTSC veto was looming over G2 as a real possibility, he deigned to provide them with the information already sent to Menzies, to avert further tensions. He summed up by saying that the 'Safety Committee is on our side, but a bit jittery'.[57]

Wary of the AWTSC veto, Penney wrote to Adams on 10 May 1956. Adams was aboard *Narvik*, preparing for G1. Penney wrote:

> Strongly advise not showing the Safety Committee any significant weapon details ... They could be told that fissile material is at centre of large ball of high explosives ... No details of explosive configuration or inner components must be revealed. Appreciate that the position is awkward for you and you must make minor concessions.[58]

Royal Commissioner Jim McClelland was scathing of this message in his 1985 report, saying that 'the British perception of concessions was that they should be very minor indeed'.[59]

By the time of Mosaic, a groundswell of public concern had begun about the seemingly open-ended program of British atomic tests in Australia.

Protests were being held in cities including Perth and some placards held by the protesters suggested that Australians had become 'guinea pigs'.[60] The leader of the opposition, HV 'Doc' Evatt, joined the chorus of disapproval as bipartisan support for atomic weapons tests in Australia collapsed. A week after G2, the Australian Labor Party caucus voted to ban British atomic tests in Australia. Evatt was quoted in the left-wing newspaper the Sydney *Tribune* as saying that a recent public opinion survey 'shows that a large majority of Australians oppose the forthcoming nuclear weapon tests at Monte Bello'.[61] Evatt was also quoted as saying, 'Australian public opinion is far in advance of that of the Federal Government which meekly acquiesces in these highly erratic hydrogen and atomic bomb experiments that are taking place with increasing frequency'.[62] In June 1956, the ALP caucus voted to oppose future atomic tests in Australia. The Australian media reported that political pressure was escalating as the ALP moved to an anti-nuclear weapons policy stance. The withdrawal of opposition support featured extensively in the media. The *Canberra Times* covered the policy departure:

> The next Federal Labour [sic] Government would vote no money for tests of nuclear weapons or the development of means of waging nuclear warfare, the Deputy Leader of the Federal Opposition, Mr Calwell, said today. Mr Calwell said that developments after the [Mosaic G2] Monte Bello atomic explosion last week [19 June 1956] showed that if nuclear weapons tests were permitted to continue there, radio active dust, despite all precautions, might be carried across Australia.[63]

The real turning point in both public and government sentiment in Australia was Mosaic G2. This huge explosion, feared by some to have been a nuclear accident, set off behind-the-scenes scrambling by both governments. Flights out of Perth were grounded for a short while, as governments reckoned with 'Australia's first radiological scare'.[64] The *Argus* newspaper reported that '[p]art or all of the cloud was picked up by a westerly wind that blew up after the explosion, and has been swept across this coast. This was obvious to mainland observers soon after the bomb was exploded'.[65] The British High Commission reported to London that the 'Australian Government have been embarrassed overnight by press story that atomic cloud drifted inland after explosion'.[66] Minister for Supply Beale was forced to issue a media statement at midnight, after receiving comforting words from (somewhat annoyed) members of the AWTSC aboard *Narvik*.[67] Beale was visiting the new Maralinga site with a group of press representatives when news broke, and he spent hours checking and preparing the statement, claiming there was no danger. The Safety Committee could have done without what they considered unnecessary fuss by the government.[68]

Bold headlines dominated the front page of the Sydney *Sun* on 21 June 1956, as disquiet about G2 spread. 'THREAT TO 3 TOWNS: Atom cloud drift'. The story began: 'All Australia is anxiously watching a radio-active cloud – result of the atomic blast on the Monte Bello islands on Tuesday'.[69] The story reported that the 47 000-foot-high cloud 'could be carried along the strip of coast through Broome, Wyndham and Darwin'.[70]

Beale's discomfort was closely monitored by the British high commissioner, who referred to it in a telegram sent to the CRO on 21 June.[71] The telegram noted that the Bureau of Meteorology in Melbourne had sent a report to *Narvik* that apparently confirmed the media rumour of mainland contamination, although this was negated by AWTSC. 'Chairman of Australian Safety Committee in NARVIK subsequently sent message to Beale that cloud was safely over sea 100 miles from land that all safety precautions had been taken and that there had never been any danger for mainland.'[72] The matter was discussed in Cabinet on 20 June 1956, and Cabinet subsequently asked the Prime Minister's Department to seek clarity on what happened.[73] Ultimately the Australian government was placated, but some trust was eroded between the two governments. The report of the Royal Commission, when examining the spike in panic caused in Australia by the unexpectedly large Mosaic G2 explosion, noted that it was 'quite apparent that, from the Australian and British point of view, press and public statements about the tests were to be made in the most comforting terms'.[74] However, the story ran in the media for weeks.

The usual blithe assurances from the federal government fell short of reassuring the opposition when they were aired in parliament two days after G2. Doc Evatt asked Fadden, as tensions rose over the latest Mosaic test: 'Can the Acting Prime Minister give the House a statement now, and later a more complete statement, regarding the effect on the Australian continent of the nuclear experiment that was carried out yesterday in the area of the Monte Bello Islands?'[75] He elaborated, remarking on the 'general concern about the matter of nuclear experiments, which has been increased by dogmatic statements and assurances given over and over again by the Minister for Supply [Beale], which assurances appear not to have been justified in the light of subsequent events'.[76] Fadden replied with the usual Australian government line, that 'the most rigid safety standards' had been observed.[77] Nevertheless, there was a hint of government concern about the aftermath of G2 as he continued his answer:

> However, in view of the reports about the existence of cloud over the mainland, the Government sought further information from the safety committee. Early this morning ... a message was received from the committee. The committee has stated that it cannot overemphasize the fact that the whole operation was carried out without risk to life or property on the mainland or elsewhere, and that there is absolutely no danger to persons on the mainland.[78]

Evatt persisted, putting the issue squarely on Fadden's shoulders: 'I ask the Acting Prime Minister to treat this matter as one for which he must be personally responsible'.[79] Fadden declared that he had sought further information, despite the assurances of the Safety Committee: 'I am in complete agreement with the right honourable gentleman that the whole matter is worthy of the fullest report, and I will see that I get it'.[80]

A briefing document prepared by the Department of Supply tracked media reaction to the British atomic tests. The change in public sentiment caused by the fate of the *Fukuryu Maru*, a Japanese tuna fishing vessel that steamed into the path of fallout from the 15-megaton US H-bomb test Castle Bravo in the Pacific, was noted as a turning point in 1954, although no atomic tests were held in Australia that year.[81] With a distinct flavour of the times, the briefing note discussed the rising influence of movements against atomic weaponry:

> [t]he international Peace Campaign, although an obvious instrument of Communism was fostered during 1954 and 1955 by people of many shades of thought who could be attracted to an ideal which was humanitarian in itself. The Press was not slow to sense this vocal pressure and, chameleon-like, began to offset any articles showing pride in British technical advances with far more attention to the dangers, both political and physical, implicit in atomic trials.[82]

The disquiet from Mosaic G2 was dismissed as 'near hysteria built on flimsy misconceptions'.[83] In public, the Australian government acknowledged the change in sentiment. Beale stated in a press release that discussed why Australia was hosting Britain's tests that

> until 1954 tests were carried out in Australia without much comment (except from the Communists); indeed, there was some general pride that Australia was able to make so valuable a contribution to British Commonwealth defence. Since then, however, uneasiness has been publicly expressed as to the dangers of atomic radiation resulting from A-tests.[84]

Charles Adams and Hugh Martell visited the British High Commission in Canberra in July 1956 to discuss the aftermath of G2 with Deputy High Commissioner Geofroy Tory and one of his colleagues. The two senior Mosaic officials were unhappy that ignorance of atomic testing among both politicians and the general public had made their job more difficult, as outlined in Tory's minutes from the meeting.

> We made it clear that the scientific assurances which Ministers and officials had received had satisfied them, but that there was still a feeling that the public should have been told before the test took place that there would be some activity overland [sic], but that it would be entirely harmless, so that they could have been spared the alarm which was undoubtedly caused when the first reports of the drifting cloud were received.[85]

They discussed a report in the *Sydney Morning Herald* of 27 June 1956 that had criticised the delay in providing information to the public as fears had grown about radioactivity crossing onto the mainland. Tory claimed that all of the blame belonged with the AWTSC as 'it was they who should have advised the Government on necessary publicity. The Australians still appeared to feel, however, that it had been up to us to alert them on this point'.[86] According to Tory's notes, Adams affirmed that the 'much vaunted' Safety Committee had access to all relevant information about the tests and their likely aftermath, which 'had been issued in unclassified form in the U.K. but the Australians for their own purposes had classified the document in question, presumably with a view to avoiding unnecessary public alarm in so far as the document made clear the inevitability of some activity over land'.[87] These four British officials, sitting comfortably together only a short distance from Parliament House in Canberra, found fault with this approach, with Tory noting that it was 'exactly the wrong reason because these were the facts which should have been put before the public, as hindsight now showed'.[88] Adams expressed some sympathy with the Australian prime minister, having to deal with the political consequences, although that was nothing to do with Britain.

After their meeting at the High Commission, Tory took Adams and Martell to see John Bunting, deputy secretary of the Prime Minister's Department (and later Australian high commissioner to the UK). They had hoped to see the acting prime minister, Arthur Fadden, but he was tied up with a Loans Council meeting. While Bunting toed the Australian government line that the fallout over land was unacceptable, he 'was very reasonable and made no real complaint'.[89] They then discussed how to run publicity in future, especially with Maralinga about to begin operations. Tory ended on a high note: 'Commodore Martell and Mr. Adams left after lunch fairly satisfied with their talk and reassured as to the atmosphere in the Prime Minister's Department'.[90]

The Australian government had been worried about the cost of Operation Mosaic from the start, given that it was unexpected and unbudgeted. Throughout the British tests, Australia always paid more than originally agreed, and the British found subtle ways of encouraging them to make a greater commitment. While early indications from the British were that Australia would be required to expend about one-tenth the effort put into Hurricane, this was quickly revised upwards, as discussed in the report of the working party on Operation Mosaic produced on 10 October 1955.[91] As Mosaic was on the drawing board, the British government was keen to get its Australian counterpart up to the funding starting line. The undersecretary at the Treasury, David Serpell, prepared a top-secret note in July 1955 to the Treasury representative in Australia, HL Jenkyns, laying out Australia's position.

> The Australian Prime Minister ... has said that the strain on Australia's resources will prevent them from contributing very much to 'Mosaic'. It is to be hoped that, when the Australians learn the relatively modest scale of the support proposed, they will feel able to co-operate to the extent required.[92]

Serpell was concerned that Menzies would insist that whatever the Australians provided for Mosaic would be on the basis of loan for which they would expect to be reimbursed. 'Every effort should be made to persuade the Australians not to charge for this support.'[93] The sensitivity of this question in relation to a test outside the agreed Maralinga program is evident in Serpell's advice to Jenkyns, that he was to tread a fine line in any discussion with the Australian government:

> every care should be taken to avoid a situation in which insistence on this point may adversely affect the Australian Government's attitude towards MOSAIC as a whole. Should there be any risk ... you should report back in order to give us the opportunity of deciding whether (a) to press the Australians further; (b) to pay them for the services in question; (c) to provide the services ourselves.[94]

Serpell went on to say that attempting to obtain Australian agreement to provide resources for Mosaic should not interfere with the commitments already made for Maralinga, an investment that Australia could, quite rightly, point to as a considerable financial input. Mosaic, after all, came out of left field. One potential sticking point in the negotiations concerned future liability for damages. An agreement had yet to be reached on future liability for any harm caused to people or property. 'This question did not arise in connection with "Hurricane" or "Totem" and we should not raise it in connection with MOSAIC. If it is raised by the Australians you should report back through the High Commissioner.'[95]

Much later, at the Royal Commission, William Penney would be pressed on the apparent deliberate policy to keep Australian involvement minimal and under-informed, except in the case of Ernest Titterton. Penney was recalled to the stand on 15 March 1985, weeks after his initial testimony. He had a bad cold and seemed to struggle with his answers. He was confronted with a letter from the CRO, sent after the public announcement of Operation Hurricane in March 1952, that said in part 'there will be no further pressure from the Australian side for fuller details or for information about the grounds on which [the assurance of the safety of the test] was given'.[96] Penney could not remember that correspondence, and emphasised that he was more focused on the technical side rather than the politics of the test. Counsel for the Royal Commission, Peter McClellan, suggested that this and other correspondence indicated that, at least at the governmental level, Australians were being actively excluded. Penney had to agree that in the early stages that had been the case, but during Hurricane itself the three Australian observers 'were in the control room' and therefore not actively excluded.[97] There was no Australian control room, though. The role of the Australians at both Hurricane and Mosaic was always largely for appearances.

CHAPTER 7

PIECING TOGETHER MOSAIC

> ... we need as soon as possible as many ... H-bombs of the minimum power necessary to provide, with the certainty of delivery of a small number, an effective deterrent. I interpret this as meaning that our H bombs need only be powerful enough to destroy a centre of government or industry.[1]
>
> SIR FREDERICK BRUNDRETT, BRITISH WORKING PARTY ON THE OPERATIONAL USE OF ATOMIC WEAPONS, SEPTEMBER 1954

Soon, Hurricane was a distant memory. As Britain shifted gears in an attempt to join the US and the USSR in producing hydrogen weapon technology, the restrictions on testing this much more powerful weaponry in Australia had to be managed. The information flow to the Australians was technical-sounding but vague. While the Australians did start asking questions about this mysterious and seemingly abrupt departure from the scheduled British testing program, the true nature of the Mosaic bombs was not fully disclosed to the government of the country where the tests would be held.

Operation Mosaic comprised two bombs – G1 fired on 15 May 1956 and G2 fired on 19 June 1956 – detonated from 31-metre aluminium alloy towers on two Monte Bello islands. They were both, technically, fission weapons, although nothing like them had been tested before in Australia. They were part of the preparations for testing a British H-bomb in the Pacific, and had some extra design elements that may have brought them close to the limits of the agreement with Australia not to test hydrogen weapons on Australian territory. These weapons were intended to provide clear directions for the British H-bomb program. Their technology was arguably a design dead end, since they ended up being negative experiments – they confirmed for the

British test authorities a direction they didn't want to go. That made them essential for deciding the next steps in Britain's march towards a much more powerful weapon.

The original plan for Mosaic specified that the second of the two planned blasts would be smaller than the first and that the first shot should take place no later than 1 May 1955.[2] These plans changed over time, as the team at Aldermaston worked against the clock on finalising weapon tests that would help to enable the building of a British hydrogen bomb, while trying to prepare a site on the other side of the world. HMS *Narvik*, which had been used at Hurricane, was in 1955 'mothballed' in the Clyde River.[3] *Narvik* had what was then known as 'coon-coating', a mesh and plastic covering to protect the vessel from the elements.[4] The vessel would have to be repainted and refitted to a tight timeline to be ready for another operation at the far-flung isles, this time under the command of Hugh Martell. In his Mosaic report produced later in 1956, Martell acknowledged the difficulties associated with the earlier Monte Bello campaign, saying, 'Operation Hurricane had shown that in the marriage of a military command with a Scientific Research Establishment the honeymoon period can be difficult'.[5] While *Narvik* was being prepared for Mosaic, a 'cover story' was formulated by the new Mosaic Executive, MOSEX, to head off any snooping by people who had noticed the ship in dock. The first MOSEX meeting, chaired by Sir Frederick Morgan, in July 1955, agreed that 'the ship's company and dockyard workmen in the ship could be told that the ship was commissioning for foreign service and would not leave the United Kingdom before Christmas'.[6]

While *Narvik* was to be the control ship, HMS *Alert* would accommodate most of the scientific party.[7] In addition, the Admiralty wanted an LCM (landing craft mechanical) and LST (landing ship, tank) to bring equipment and personnel onto the islands. Given that it would cost £25 000 to fit out an LST, Admiralty declined to proceed further until a budget for this cost was allocated. 'Unless this is done, no further dockyard work can be started and the dates cannot be met.'[8] So much of the history of British atomic testing is dominated by the budgetary constraints applied by the British civil service. Eventually the money was found.

Operation Mosaic echoed its codename in being an exercise in piecing together different technologies to create a new kind of weapon. This was probably why it replaced the original codename, Operation Giraffe.[9] The 'G' for Giraffe lingers on in the device codenames G1 and G2. While the atomic test code names were not generally known in public at the time, one British journalist did obtain this name, and speculated in print about what it might mean. Denis Warner, who was in the press party for G1, wrote in a 1956 story in the UK's *Daily Telegraph*:

> Only in the unannounced code name for the test, 'Operation Mosaic', was there an official clue that this experiment was part of a much wider field of research and development. Since the code name was not a military inspiration, and in view of the earlier hints, it is a safe assumption that the bomb whose shadow hangs stupendously above this minesweeper as I am writing this was either a trigger for a hydrogen bomb or was directly related to some other aspect of the production of a much more powerful nuclear weapon.[10]

Warner was remarkably astute, given how little information was forthcoming from the British authorities, most of whom despised and distrusted the press. Mosaic was most definitely related to the megaton weapons that William Penney and his team were striving to create at Aldermaston to a tight timeline for testing in the Pacific in 1957.

Britain did not want to wait until the Maralinga range was ready in September 1956, since the R&D around the H-bomb was proceeding apace, and a moratorium on atmospheric testing was imminent. Also, arrangements for Operation Buffalo were well advanced and would need to go ahead at Maralinga as planned in September–October 1956. Perhaps more compellingly, though, Penney did not believe that Australia would agree to test at Maralinga what they had in mind for at least one of the Mosaic bombs. This was a weapon that they knew would be of huge yield, because the British scientists were examining a potential alternative to a true hydrogen bomb. Penney's offsider, William Cook, was running the AWRE weapons development committee, and the name of the game by then was thermonuclear. As Penney later said in evidence at the Royal Commission, 'Maralinga was not going to be possible; it was too early; and if we had said to the Australians 50 kilotons at Maralinga I think they would have said no. So we could not go there'.[11] The other option for the nascent British high yield weapon was to try and partner up again with the Americans, but as Penney said, '[w]ell, we had been through that hoop'.[12] Monte Bello it had to be, or there would be no Pacific H-bomb test for the British in 1957. The weather would only be right for one day in 100, according to the Royal Navy meteorologist Richard (Dick) Fotheringham, so they would be threading a needle. Planning for Hurricane had shown that only October was a suitable month for testing, not May or June. The British penchant for pushing the testing envelope was clearly evident.

At the very beginning, Mosaic had been planned as a single explosion, but in April 1955, 'for reasons of scientific policy', it was changed to add a second device.[13] The scope of the project expanded greatly from its original size. As it grew, the British government created the Atomic Weapons Trials Executive, chaired by the controller of atomic energy, Sir Frederick Morgan, to become the authority dealing with all aspects of the operation.[14] Soon after, Morgan was appointed chair of the MOSEX, a group that had members from the ministries of Supply and Defence, the CRO, AWRE, the Admiralty, Air Ministry and War Office.[15] Work was soon underway to develop protocols and experiments to make the most of the opportunities afforded by this new Monte Bello series.

In May 1955, the UK government asked the Australian government for permission to conduct more atomic tests in the Monte Bello Islands, initially indicating they would take place during April/May 1956.[16] The Australian government replied on 20 June 1955, agreeing in principle to host the tests at Monte Bello, and later asking that UK representatives visit Australia to discuss details, and to give particulars of the support required from Australian resources.[17] This request for more detailed information was new, and behind the scenes caused some consternation in Britain. In July 1955, the minister for defence, Philip McBride, wrote to the minister for the navy, Josiah Francis, to point out that the Defence Committee believed the diversion of resources would 'adversely affect other commitments and could only be undertaken if high priority were accorded to the project by the Government'.[18] The Australian government had already undertaken to provide 127 military personnel to help build the firing ranges at the new

Maralinga test site, from March 1956.[19] Nevertheless, somewhat grudging approval was given to the British proposal.

MOSEX held its first meeting on 6 July 1955, chaired by Sir Frederick Morgan. Task Force Commander Hugh Martell was one of the attendees, as was Penney's deputy, Charles Adams. The small agenda covered preliminary planning for the operation. The assistant deputy director of AWRE, Admiral PWB Brooking, noted that 'there would be no security requirements in Monte Bello subsequent to the operation. A.W.R.E. would inform the Australians of the extent of the contaminated area left there'.[20] The meeting discussed the request from Robert Menzies to send a party to Australia for meetings on the details of the operation.[21] They agreed that, because trips by British officials to Maralinga and Woomera had become so commonplace, there was 'little security risk' in this visit, made easier by the fact that the group would not be visiting Monte Bello itself.[22] Accordingly, a party from the UK conferred with a sub-committee of the Maralinga Committee on 26, 27 and 28 July 1955.[23] This party comprised Captain Hugh Martell from the Royal Navy, who would become the Mosaic task force commander, Group Captain Paddy Menaul from the RAF, who would become the air task group commander for Mosaic (and fly a Canberra aircraft through the G1 cloud), Lieutenant Commander AK Dodds, staff officer to Martell, and meteorologist Lieutenant Commander Dick Fotheringham, who was also representing William Penney.[24] Owing to a problem with their aircraft in Singapore, their time in Melbourne was reduced to just a few days.[25] They met with a group of senior Australian bureaucrats and military personnel chaired by the secretary of the Department of Supply, Frank O'Connor, including Captain William Marks from the RAN (later chair of the Australian committee that would manage Mosaic), Air Vice Marshal William Hely from the RAAF, HC Newman from the Department of the Treasury, EL White from the Department of Defence, and several senior representatives from the Department of Supply.[26] This was the first opportunity for the British and Australian authorities to talk in detail about what Mosaic would entail. Specifically, the Brits wanted to sound out the Australians about the level of support Australia was willing to provide for the operation, given the hints at some constraints in Menzies' communication with Eden.[27] The two groups agreed that while Britain would cover the cost of the operation itself, Australia would contribute the expenses of any Australian personnel authorised at the request of the Australian government to attend the trials, and the cost of pay, allowances, maintenance and transport of Australian service personnel provided for agreed tasks.[28] The Australian government would also offer a range of administrative assistance. During 1955, the Australian government had to revise upwards the commitment it had signed up to. The report of the Montebello Working Party noted in October that while it had been thought that the effort required would be about one-tenth of that required for Hurricane, 'later considerations indicate that the effort involved may well be much greater'.[29]

ASIO agent Leo Carter, who had been involved in the Hurricane project, was brought into Mosaic planning.[30] At the first meeting of the working party, held in Melbourne on 14 September 1955, Carter told the other members that ASIO would work with the military to meet all security requirements. 'The Australian Government would bear the cost of Security arrangements.'[31] At the same meeting, Carter referred to the need for making suitable arrangements for dealing with journalists, and stated that while he as a security officer could not act as a press liaison officer 'for obvious reasons' he would be willing to advise on press releases. Commander GL Fowle, secretary to the committee, backed him on this, saying, 'that the records of the previous test showed weaknesses in the handling of press liaison'.[32] No member of the committee volunteered to help out with media liaison, so the matter was deferred.

After Mosaic (and Operation Buffalo at Maralinga in September 1956), the secretary of the Department of Supply, Frank O'Connor, wrote to the head of ASIO, Brigadier Charles Spry, thanking him for making Leo Carter available. He said that it 'was largely due to his untiring efforts and willingness to undertake any task that the security service provided for both operations met with the approbation of all United Kingdom and Australian Departments and Authorities concerned'.[33] Carter also met with the British delegation to discuss security. He suggested that he should be on board *Narvik* for the operation, and the delegation agreed.[34] Charles Adams met the next day with Alan Butement, representing the new AWTSC, and was given a broad description of the operation, including the anticipated yields.[35] Butement raised the intriguing issue of information that might be obtainable by remote sampling of the cloud, presumably undertaken by the Soviets as a way of monitoring the tests in Australia. According to a MOSEX briefing document, 'he wished to have some statement from U.K. on this question since, from a political point of view, it would be important to Australia'.[36] Adams suggested that this issue could be taken up at a further meeting in Britain in early 1956. William Penney would be asked at the Royal Commission whether it was possible that the Soviets had monitored the tests in Australia by air sampling, and revealed that it was.[37]

Australia made its first major commitment to Operation Mosaic in September 1955 when it established the Monte Bello Working Party, after initially placing planning into the larger Maralinga Sub-Committee within the Department of Supply.[38] This working party, chaired by William Marks,[39] would co-ordinate the Australian end of the Mosaic project. Other members were Commander GH Blaxland from the UK Ministry of Supply Staff (Australia) UKMOSS(A), H Loxton from the Australian Department of Supply and Commander GL Fowle, secretary to the committee.[40] Leo Carter was present at every meeting, although his name was usually redacted in the minutes. Various others attended as required, including representatives from the RAAF, the Royal Navy and the Bureau of Meteorology. In October 1955, the same ships that had provided services before Operation Hurricane, HMAS *Warrego* and HMAS *Karangi*, were pressed into service again at Monte Bello, undertaking a check survey and re-laying buoys to mark channels.[41]

That the Australians were taken aback by the decision of the British to return to Monte Bello is signalled by the fact that Mosaic planning initially had to fit in with the Maralinga planning operations already on foot. The Australians had thought that Maralinga was the next stage in the development of British nuclear weapons. The sudden return to the old testing ground meant that compromises had to be made. Secretary of the Department of Supply Frank O'Connor was most displeased at this turn of events. In a top-secret briefing document for his minister, Howard Beale, he said that they had their 'hands full with Maralinga, and we are certainly not looking for further work'.[42] The Australians had also received no information at all about the 'new dimension in the design of the devices', the inclusion of light elements, that had only been disclosed in outline to the Australian government via Anthony Eden's initial message to Robert Menzies.[43] The Royal Commission would later find 'for Mosaic, they were unable to check the calculations of probable yield. The speed with which the decision to set up Mosaic and fire the weapons gave the Safety Committee little time to prepare itself properly in advance'.[44] The exact nature of Mosaic was by no means clear to the Australian government.

AWRE officials held extensive discussions about how much technical detail could be provided to the Australians. They were bound by their security agreements with America and could not release any information that touched upon any part of the technology that had originated in the States. A minute paper between EP Hicks and AWRE Assistant Deputy Director Brooking sketched out what might be permissible to include in material sent to Australia:

1 The reports do not reveal, either directly or indirectly, the efficiency or internal design of any atomic weapon, and
2 The reports do not contain any classified information of U.S. origin, except that certain U.S. military Classified information is releasable to Australia by agreement of the United States Government.[45]

In effect, AWRE agreed that Australian officials could have access to non-diagnostic information on the effects of atomic weapons obtained in British weapon trials, including material that might hint at the yield of weapons exploded on Australian territory, and including details of the instrumentation used in the measurements.[46]

Charles Adams sent initial briefing material to the AWTSC in September 1955, via the UKMOSS(A) office in Melbourne. This material indicated that AWRE would base its predictions of contamination and cloud behaviour for Mosaic on a theoretical model developed using data gathered from Operation Totem at Emu Field in 1953.[47] AWRE meteorologist Albert Matthewman headed the technical predictions group at Aldermaston, charged with making detailed forecasts of atmospheric conditions for Mosaic. The group's work included predicting the height of the atomic cloud, vital to understanding where fallout might occur. Cloud height depends on weapon yield and ambient atmospheric temperature and humidity.[48]

Despite the work done on this at Aldermaston, the predictions for G1 were seriously underestimated. The G1 cloud was predicted to rise to 17 500 feet but actually rose to 24 000 feet.[49]

Meteorological and cloud predictions also drew upon weather data from the director of the Commonwealth Bureau of Meteorology, Leonard Dwyer, who would become a member of the AWTSC from 1957. Dwyer indicated a narrow range of options were available in the timeframe to minimise the risk of the cloud moving over the Australian mainland. The information provided to the AWTSC mentions the possible height of the cloud expected for G2, suggesting planning for anything up to 50 000 feet,[50] well above the Totem clouds (Totem 1 went to 15 000 feet and Totem 2 to between 20 000 and 28 000 feet), which were much smaller bombs (9 and 7 kilotons respectively). The higher the yield of a nuclear weapon, the higher the cloud. In the event, the G2 cloud rose 12 000 feet higher than predicted, to 47 000 feet.[51]

Australia contributed substantial resources to Mosaic. A ground survey involving a four-week deployment of HMAS *Warrego* to the islands got underway in early October 1955,[52] primarily to erect navigational marks, and lay buoys and moorings, and to advise on whether landings could be made on Trimouille Island, proposed site of G1. In addition, Australia undertook to provide an ocean mine sweeper for ferry duties between the islands and Onslow, as well as a motor water lighter and a motor refrigerator lighter.[53] Arrangements were made with the WA State Shipping Service to charter a ship for transporting stores from Fremantle to Onslow, and, in some cases, out to the Monte Bellos.

The RAN established a shore wireless station at Onslow, with a small working party, and was responsible for providing food and other provisions for all Australian naval vessels involved.[54] The RAAF supplied airfield and accommodation facilities, decontamination facilities, weather forecasting, Neptune aircraft for security patrols, air/sea rescue, and general domestic facilities for the RAF component.[55] The list of requirements seemed endless and kept growing.

The operation was going to generate significant contaminated waste. The RAAF had said from the start that it was not prepared to deal with it, especially in light of the toxic mess left behind from Hurricane. This issue was discussed at most of the Monte Bello Working Party meetings. Advice was sought from the UK, and at the January 1956 meeting, advice from Paddy Menaul was tabled. He confirmed information from AWRE that only partial decontamination would take place at Pearce, and major decontamination at Amberley, near Brisbane. All low-level liquid effluent would be allowed to soak away into the ground at Pearce, while low-level solid waste would be sealed in drums. Active liquids would be turned into sludge and placed into drums for disposal at sea. More highly contaminated active waste was to be shipped to Maralinga for disposal.[56] At the following meeting, attended by both Charles Adams and Ieuan Maddock, the proposal for soaking low-level waste at Pearce was rejected, and a lively discussion ensued.[57] The matter was not resolved, although the RAAF representatives were adamant that assurances about the safety of the waste were not going to sway them.[58]

Unaware that the Australian government was sensitive about the fact that most Australians at that time identified as Christian, the UK authorities initially forged ahead with a plan to test Mosaic G2 on a Sunday, 10 June. Australian Minister for Supply Howard Beale, who was responsible for the Australian end of the operation, was horrified when word came to him of the proposed date. The CRO notified the UK High Commission in Australia that a new date was needed; Sundays were out of the question. Beale was immovable on this topic, which would affect firing dates at Maralinga as well. The CRO noted that the 'objection is on religious grounds and Mr. Beale feels he would have no adequate answer to the strong protests which would arise. The Operational Commander in deference to this view was instructed that firing must not take place on Sunday 10th June'.[59]

While the UK made the change to the G2 firing date, the authorities were far from happy. The UK minister for supply, Reginald Maudling, was concerned at the effect that prohibition of Sunday firing would have on the Mosaic test schedule if strictly enforced.[60] 'Opportunities for firing the second round are likely to be very limited and the Force may therefore have to wait some days or even weeks before it is safe to fire.'[61] The British wanted this operation wrapped up as quickly as possible, not least because the lead naval vessel HMS *Narvik* had to be re-equipped quickly to be ready for Operation Grapple at Malden and Christmas islands in 1957.

> From a purely operational point of view it is therefore undesirable that the Commander [Martell] should be inhibited from firing on any day on which meteorological conditions are right. He has himself cabled 'Opportunities for firing G2 are likely to be very rare if at all', and he says he would deplore a decision not to fire on Sundays.[62]

As a result of British government disquiet about the Sunday rule, the CRO asked that Beale be told that if it had to be a Sunday, then so be it: 'Please inform Mr. Beale that to meet him we should be ready to instruct Operational Commander not to plan deliberately to fire on a Sunday but on strong practical grounds seek his concurrence that if next suitable day should by chance fall upon Sunday firing will not on that account be postponed'.[63] A holy battle over a Sunday firing did not eventuate, however, as Tuesday 19 June ended up being deemed suitable.

The timing of the Mosaic blasts, over and above the avoidance of Sundays, was a headache for the British authorities and the subject of much back-and-forth over the telegraph wires. Originally, the latest date for G1 was planned as 1 May 1956.[64] However, various delays complicated planning. A cyclone had passed over the area in February, pushing back the arrival of HMS *Narvik* by five days. This delay, combined with a severe shortage of manpower, meant that crew from HMS *Alert* were pressed into working onshore to try and prevent the timetable from drifting further.[65]

Everything had to stop, though, for a VIP. On Sunday 15 April, the head of the Royal Navy, Admiral Earl Louis Mountbatten of Burma, First Sea Lord, second cousin to Queen Elizabeth, visited the Monte Bellos. He arrived by helicopter and was shown around the weapon and control sites, later addressing the ship's companies of HMS *Alert*, HMS *Narvik* and HMAS *Fremantle* on Alpha Island,[66] where Mosaic G2 was to be detonated. Mountbatten had visited *Narvik* as she was being fitted out for Mosaic at the Portsmouth Dockyard just before sailing to Australia the previous Christmas. Hugh Martell hosted a buffet lunch of 'Monte Bello chicken and Geraldton crayfish'[67] for the aristocratic visitor, before Mountbatten returned to Onslow. While the Sea Lord's whirlwind trip to Monte Bello was not publicised, the next day in a public address he discussed the atomic future of the Royal Navy, predicting that 'nuclear energy would become the main source of propulsion for both naval and merchant ships',[68] with nuclear power for submarines being the first step. On the same trip, he also visited Woomera. Mountbatten held the role of First Sea Lord from 1955 to 1959, modernising the fighting service during his tenure. He was to play a significant role in British acquisition in 1968 of its submarine-based nuclear deterrent, Polaris, based on an American design, ensuring an ongoing role for the Royal Navy in the country's nuclear arsenal.[69] Mountbatten would, in 1979, be killed by an Irish Republican Army bomb.

Hugh Martell looms, larger than life, over Operation Mosaic. The commander of Task Force 308, father of 14 children, was remembered in one of his obituaries for 'the sheer force of his personality'.[70] At the Royal Commission, Martell played down his central role in the conduct of Mosaic, claiming to have at best minimal knowledge about the nature of the tests, on a need-to-know basis only. 'Remember that I am a naval officer, not a scientist or even a mathematician. I have to rely on the detailed advice, scientific and technical, of my staff, whether they are scientists or naval people, or airmen or what have you.'[71] He continued:

> I was given no precise details of these tests, what they were for or why, or anything. All I can say is that one got to understand certain facets of these tests because on board *Narvik* ... the Scientific Director, Mr [Adams] ... and I every two days had lunch fortified by red Australian wine, and we got to learn a lot of each other's business.[72]

When asked whether we would have been told that the explosions would differ from conventional atomic bombs through the addition of light elements he replied: 'No, I had no need to know that sort of fact'.[73] At the time of Mosaic, though, Martell was disposed to throw his weight around a lot more than he was willing to admit at the Royal Commission. In his Operation Mosaic report, he sets out his central role, including making crucial decisions about how the scientific and military sides would work together, decreeing that Aldermaston was to be the centre of operations.[74] He claimed credit for enabling close co-operation between the scientific superintendent and his naval operations staff officer. He was also entrusted with leading the team sent to Australia to discuss the operation with the Australian government. He knew a lot about it. The Australians understood this too – the minutes of the third meeting of the Monte Bello Working Party held on Friday 30 September 1955 noted 'that the Task Force Commander would be the officer responsible for deciding whether or not the tests would take place'.[75]

As Mosaic was gearing up, the British government prepared carefully worded publicity material for the media. As part of this, the sort of questions likely to be asked were forecast and answers formulated. At the top of the list was the likelihood that a connection would be made between Mosaic (and Buffalo) and the British H-bomb announced in February 1955. A top-secret document made this suggestion about the H-bomb speculation: that saying 'no' was the safest answer, but the preference was to say that for security reasons information about the precise purpose of Mosaic could not be supplied, other than to say that no hydrogen weapon would be tested in Australia. The note went on to say 'sooner or later the connection of the tests with the development of the megaton bomb may become known and if we give a direct negative we could be accused of misleading not only our own public but also the Australians'.[76] Most of the foreshadowed questions, such as arrangements for press, were to be answered with a bland 'this is under consideration'.[77] However, the process of placating was firmly established in relation to the putative question, 'how will these tests compare with the previous tests?', to which the answer was given: 'they would be approximately of the same order'.[78]

The newly formed AWTSC needed more information about Mosaic, and particularly its 'light elements', which set it apart from previous tests in Australia. Britain did not initially wish to furnish this information, preferring to keep the characteristics of the Mosaic bombs vague. However, when the AWTSC indicated that a lack of information might lead to the deployment of its veto, Adams and Martell sought to avert a deadlock.[79] After obtaining assurances from Leslie Martin and his committee that the information would be kept secret, further detail was provided to the Australians. While billeted on HMS *Alert*, the AWTSC joined daily meetings on *Narvik*. When on board *Narvik*, Martin and his committee sought the allocation of a private room to consider the information, away from Martell and Adams. He wrote to the chair of the Australian Mosaic committee, Captain Marks, saying 'we fear that our interests may clash with those of the UK team'.[80] As Symonds remarked, 'no mention of the content nor the interpretation placed by the AWTSC on the information given appears in Australian documents which have been made available. There are no records of any meetings held by the AWTSC while they were on NARVIK'.[81] Nevertheless, it seems likely that the committee was provided with some technical detail about Mosaic, since they provided no impediment to it proceeding.

Both the British and Australian authorities had learned the benefits of managing the press and feeding through positive information about the tests. In preparation for Mosaic, extensive plans were enacted for media materials. Media liaison was to be co-ordinated by the Department of Supply, working with naval public relations.[82] An early draft of a press announcement that mentioned both Mosaic and the coming trials at Maralinga later in 1956 contained this line about the Maralinga tests: 'No tests will exceed a few tens of kilotons in yield and some will be smaller'.[83] This had to be struck out. A top-secret memorandum between Allen Brown in the Prime Minister's Department and Frederick Shedden in the Department of Defence explained why: 'we doubt the wisdom of including a reference to the yield of the Maralinga tests. By including this and making no reference to the Mosaic yield undesirable speculation as to the latter might well be stimulated'.[84] The memo alluded to 'some special features concerning the nature and yield of the Mosaic test'.[85] Brown also mused about whether it was wise to put out the announcement before the Australian government had been provided with all the safety assurances it was seeking, or whether the draft's general statement of safety 'was thought adequate'.[86] An amended version of the announcement was released in Australia on 12 September 1955, with no reference to the expected yields of any of the tests.

Even so, the Australian media did start to speculate whether the Mosaic tests involved hydrogen weapons. One story, in the Sydney *Sun* on 13 September 1955, quoted Mark Oliphant as saying 'that at the most the test could be of a trigger mechanism for an H-bomb',[87] reasonable speculation at the time. To head off conjecture, the Prime Minister's Department prepared a brief statement on the nature of the Mosaic weapons, echoing Eden's initial approach about the inclusion of light elements as a boost. The departmental briefing notes stated that the smaller device was 'a fission weapon which in order to get scientific data contains small quantities of the light elements used in thermonuclear bombs. The light elements are expected to react but the quantities are so small that the effect on the yield is small'.[88] The larger Mosaic device 'is a fission weapon containing somewhat larger quantities of the light elements but containing no uranium with which the light elements can react. The behaviour of the light elements will be diagnosed with the help of traces of various materials'.[89] Later, despite this announcement, uranium would be added to the design.

Secret documents were exchanged between British authorities as the country tried to ensure the messaging was finely tuned and the Australians were on board. A telegram from the CRO to the acting UK high commissioner in Canberra, Geofroy Tory, stated that

> we are arranging that United Kingdom Government Departments in dealing with any questions arising out of preparations that they are making will be careful to avoid giving any indication that preparations, e.g. commissioning of ships, are linked with further series of trials. We hope very much that the Australian Government will agree to take same line until statement is agreed with them.[90]

However, the time this message was sent, July 1955, was too early to inform the public of the planned Monte Bello tests, as the exact wording needed to be refined and ministers fully briefed. 'We understand that this would be agreeable to [Australia's Minister for Supply] Mr. Beale, who is anxious to be in Australia when first announcements are made.'[91] Public information would have to wait until September of that year.

In January 1956, the British government followed up with further public information. The announcement was drafted during December 1955 and January 1956, undergoing a typically rigorous vetting process. This communication outlined the air force involvement in Mosaic, under the command of Group Captain Menaul, holder of the Distinguished Flying Cross and a member of RAF Bomber Command during the Second World War. More than 20 RAF aircraft, plus a number from the RAAF, were to be involved in the operation.[92] They would operate out of Pearce Air

Force Base (not Geraldton as originally planned) and Onslow. (Pearce was deemed more suitable than Geraldton because of the meteorological facilities there.)[93] In parallel, an announcement was prepared outlining naval involvement in Mosaic. HMS *Narvik* was named as the lead ship. The eight ships of Task Force 308, which included one that would become known as the Mosaic 'guinea pig' ship, HMS *Diana*, would be commanded by Admiral Hugh Martell. According to a CRO briefing document, 'for this operation NARVIK has specially fitted out at Chatham, her home port. She has been at Portsmouth since early December making final preparations for her voyage to Antipodes'.[94] Martell arrived at Fremantle aboard *Narvik* on 23 February 1956.[95]

Media were alerted in April 1956 that the Monte Bello area would become hazardous to shipping and civilian aircraft from 1 May. A message went out from the Department of Civil Aviation saying that from that date 'no aircraft will fly within a radius of 150 miles or above 8000 ft. within a radius of 600 miles from Flag Island unless specifically cleared by the Air Traffic Control, Perth'.[96] A similar warning was issued for ships, in line with the prohibited area first established for Operation Hurricane as specified by the *Defence (Special Undertakings) Act* of 1952. A specific declaration for Operation Mosaic had been made under the Act in mid-February 1956, and sea and air patrols of the area began.[97] A scaled-back rehearsal for G1, using the codename Glimmer Charlie, was conducted on 5 May. The Canberra aircraft were not included because they had been grounded by bad weather. Their rehearsal took place on 8 May.[98]

Among Martell's fleet were five ships that would alternate as the designated weather ship, on duty from 9 April 1956 to provide continuous weather information.[99] In 1956, no weather satellites existed and a variety of methods to measure of atmospheric conditions were used. Accurately tracking the development of storms and cyclones in remote areas was especially important, but not always reliable. The ships that shared the weather ship role during the operation were HMS *Cossack*, HMS *Concord*, HMS *Comus*, HMS *Consort* and HMS *Newfoundland*. They operated well out in the southern Indian Ocean, close to Heard and McDonald islands. The first on duty was *Concord*, and a meteorological officer joined the ship in Singapore, and later moved between the other designated weather ships. A rating on board the ship was trained to fill weather balloons that would track the wind movement and speed of various atmospheric layers, and otherwise assist the met officer. These balloons were tracked by radar operators to gauge speed and direction of the upper winds. The Operation Mosaic trial orders emphasised that efforts should be made 'to observe winds to as great a height as possible, the results being of the utmost operational importance'.[100] Later, the AWTSC would laud the work of the Mosaic weather ships, saying the crew 'had spent monotonous days measuring the winds blowing in the upper reaches of the atmosphere and making weather reports'.[101] In particular, the ships were collecting data on the upper wind structure and the rate at which the temperature and humidity decreases with altitude. The data from the ships was augmented by readings from an RAF squadron flying weather reconnaissance missions hundreds of kilometres west out of Darwin, searching for storms and bringing back atmospheric data from areas where no ships were able to go. While ideally two flights per day were needed, only one was deemed practicable by the RAF and this compromise was accepted by the British authorities and the AWTSC.[102] Weather stations from the equator to Cocos Islands, New Zealand and the Antarctic completed the meteorological jigsaw, generating a huge volume of data that had to be dealt with using relatively primitive systems. Information was compiled on board *Narvik* to provide the task force commander with a picture of weather conditions as comprehensive as possible to enable 48 hours' notice of a suitable window of opportunity for testing. The AWTSC commented in its statement the incoming reports were 'systematically analysed hour by hour throughout the day to enable a prediction to be made of the day on which suitable firing conditions would be likely to occur'.[103] The AWTSC was acutely aware that the May–June timeframe could abound with unsatisfactory weather conditions. Winds at this time were likely to be directed towards the mainland for much of the time. Also, while the cyclone season was largely over by the end of March, the months-long preparations could still be impacted.[104]

The RN's Dick Fotheringham was the meteorologist-in-charge for Mosaic. His job was to plan the forecasting operation and ensure the right equipment was available. He worked with two Australian meteorologists, Henry Phillpot and Robert (Bob) Southern, who provided liaison with the AWTSC.[105] Among his many tasks, Fotheringham set up an upper air temperature and wind-finding station at the southern end of Hermite Island. The station was staffed by two naval ratings who launched weather balloons and tracked them by radar.[106]

In evidence to the Royal Commission nearly 30 years later, Fotheringham spoke of how he worked at Mosaic with the two Australians. 'Making a meteorological forecast is really a one-man business since otherwise you end up with an Irish Parliament and generally choosing the wrong option.'[107] Nevertheless, they worked harmoniously and he remained friends with them later. When they each separately prepared a dummy forecast before the first Mosaic shot to test their processes, 'we found that the differences were in fact negligible, which ... was highly satisfactory if slightly surprising'.[108] One of the difficulties they faced was the severe lack of data available on the tropics in the 1950s. Fotheringham said in evidence that the problems faced by the met team would one day become obsolete,

> largely overcome by the availability of satellite imagery which now has removed many of the problems which were present in the fifties. Clearly I would have liked a very close network of observation stations, but in the southern ocean and the surrounding areas of Monte Bello it was not feasible, and what we tried to do was have the necessary minimum to ensure success.[109]

When asked whether he was under pressure because of the time constraints on the firing of G2 in particular, Fotheringham responded, 'I personally was under no pressure to produce suitable weather ... if the weather did not arrive we would just go home'.[110]

The puzzle pieces of Mosaic were in place, and the military and scientific teams assembled and started converging on the Monte Bellos again. Those present would again have a front-row seat to the new world of British defence, but so much had changed: Hurricane and Mosaic were worlds apart.

CHAPTER 8
BACK TO THE ISLANDS

> The witnessing of the explosion was a rare privilege and the experience of actually seeing matter being transformed into energy will not readily be forgotten.[1]
>
> COMMANDER OF HMAS *MOTOR WATER LIGHTER 251* AT MONTE BELLO, 17 MAY 1956

> The impression of vast power being unleashed could not fail to strike everyone on board, but, due no doubt to increasing familiarity with photographs and accounts of similar explosions, the general effect was less awe inspiring than would have been the case a few years ago. The Press spent some time and typewriter ribbon trying to think up alternatives to the classic similes about mushrooms and umbrellas.[2]
>
> LIEUTENANT COMMANDER SMITH, CAPTAIN OF HMAS *JUNEE*, UPON WITNESSING MOSAIC G1

The weather was calm and hot as personnel began arriving at Monte Bello and set to work on the multitude of tasks that had been allocated to them for the coming test series. While the Mosaic Joint Operational Plan specified the clothing they should bring (windcheater, lightweight slacks, overcoat, six sets of underclothing, a 'housewife' kit to mend their clothes),[3] most of that was not necessary out at the islands. In general, they wore shorts and sandals only, sometimes complemented with straw hats and polaroid sunglasses.[4] The hard physical labour of laying the concrete foundations after the sun had gone down and erecting the massive 31-metre bomb towers was made bearable by hot meals and boats delivering tea during breaks, and 'rum rations at all hours'.[5] The towers had been manufactured back in

Britain with only minimal specifications, leading to some complications as the structures did not join together as well as they should. No one on site had specialised experience in erecting these huge towers, but Task Force Commander Hugh Martell commented that 'the good climbers were soon sorted out, and proved themselves to be fearless and satisfactory workers at heights. One of the men employed on the tower was a plumber whose hobby is mountaineering'.[6] As ever in remote and physically difficult locations, the workers found ways around the obstacles.

The islands buzzed with activity all over again, nearly four years after Hurricane. The labour was hard and around the clock. Infrastructure work required some 408 tonnes of aggregate and 90 tonnes of rapid-hardening Portland cement, and 45 tonnes of fuel in circular jerry cans, sourced from Australia.[7] A party of Royal Engineers, two officers and 66 men, arrived in February 1956 to get to work with the Australian Army to turn these materials into nuclear test facilities.[8] In addition, there were shipments from Britain of about 280 tonnes of stores, of which about 226 tonnes were for engineering purposes – towers, plant, transport and shelters.[9] The rest were about 36 tonnes of scientific stores and 18 tonnes of generators.[10] All of this gear had to be unloaded from ships rocking in the swell, then carted with limited transportation infrastructure to wherever it was required on the rocky islands.

Scientific teams started arriving in April, in levels of travel comfort depending upon seniority in the system. As the Joint Operation Plan for Operation Mosaic made clear: 'Transport of all members of the Scientific Party will be by normal [British Overseas Airways Corporation] commercial flights to Perth; Tourist Class will normally be used except where the grade of the officer entitles him to First Class'.[11] HMS *Alert* delivered the scientific crew to Monte Bello on 7 April. Several scientists, including Ieuan Maddock, had arrived a couple of days before the main party. In all, the scientists numbered 33. Mosaic scientific director Charles Adams was to have been helicoptered in on 22 April, but both aircraft were unavailable that day so he arrived by sea.[12] According to Hugh Martell, the scientific and military groups, so much at odds with each other in the lead-up to Hurricane, blended well this time on the islands. 'Full integration of scientific personnel and their organisation with that of the ships was achieved most smoothly and a spirit of co-operation was maintained at a high and happy level throughout the trial.'[13] Hard lessons had been learned from the first British test, and this time there was no Leonard Tyte or Arthur Torlesse.

Rather than control the detonation of the Mosaic bombs via radio links from *Narvik*, the electronics were instead set up on Hermite Island. The control building high on a hill that had been constructed for Hurricane was resurrected, and the firing sequence was controlled from there.[14] This measure removed some restrictions on the movement of *Narvik* and enabled the lead vessel to carry out operational duties such as meteorological analysis and control of boat movements, freed from any scientific responsibilities. About eight scientists were based on the island for the two bombs.

The fissile material for Mosaic G1 arrived at Onslow by a special RAF Hastings flight on 11 May. The yellow container that housed the explosive core of the weapon was quickly loaded aboard HMS *Alert* for the voyage out to Monte Bello.[15] Secrecy was strictly maintained, as Martell noted gleefully: 'as far as is known, Press opinion in Australia still favours the theory that [HMS] DIANA carried the vital "gimmick"'.[16] The other device components were also ferried to Gladstone Beach on Trimouille Island and assembly began in deepest secrecy. Armed sentries were on duty around the clock, and only a tiny handful of personnel were allowed anywhere near the device. The bomb was assembled and ready by 14 May to be mounted on a 31-metre tower on the north-western side of Trimouille Island, and the order was given to 'stand by'.[17] The countdown for G1 began at 11.47 am under the firing order codeword Hotshot. Commander Martell described what happened next, as the Count of Monte Bello, Ieuan Maddock, recreated his dramatic moment from three and a half years earlier:

> Mr. Maddock's voice giving the count-down was heard clearly by NARVIK, ALERT and the aircraft. Ships companies massed on deck were fallen in with their backs to the Islands, and the contrails of the sampling Canberras could be seen clearly coming in high over Barrow Island and going into their pre-arranged orbit just before H-Hour.[18]

The tension and excitement were palpable. He continued, saying that although the day was brilliantly sunny, 'the flash of the explosion was considerable'.[19] Within 20 minutes, the circling Canberra aircraft – one containing Paddy Menaul – had entered the cloud. The rest of the day was filled with activity, as ships and aircraft went through their rehearsed paces. Peace descended once evening arrived, and with it the nagging sense that the job was not complete. There was still another bomb to detonate, and it would be much bigger than G1.

While G1 did not cause the level of consternation in the Australian government or public that was to come with G2, questions were asked in parliament in its wake. On 6 June, Minister for Supply Howard Beale was asked about a report from the University of Adelaide that radioactive rain had fallen in South Australia after the first Mosaic explosion.[20] Beale denied that G1 was to blame, and indeed downplayed any risk from atmospheric testing, saying that both governments 'share the conviction that the radiation dose to human beings arising from the testing of these weapons at the present rate is insignificant compared with the radiation dose rising from natural causes'.[21] Labor member Cyril Chambers was doubtful:

> whilst the Minister's report states that no relationship can be proved between the tests and the radio activity of rain in certain places, it must be admitted that it is extraordinary that, notwithstanding the advice of the experts, rain which has fallen in South Australia has never before shown radio-activity. The matter cannot be lightly passed over, when it is only a few days or a week since the test was held.[22]

The sense of unease produced by G1 was soon to escalate when G2 exploded.

Two small, easily manoeuvrable Australian naval boats, known as *Motor Water Lighter 251* and *Motor Refrigerator Lighter 252*, made frequent runs from the mainland to Monte Bello and between the islands, transferring stores and water.[23] Both vessels were specially commissioned for Operation Mosaic.[24] The Motor Refrigerator Lighter came to be known as the 'Frigid Midget', with an unofficial logo featuring 'an Eskimo holding a carcass of beef against an Arctic background'.[25] The Motor Water Lighter, nicknamed *Gunga Din*, provided a festive atmosphere at the detonation of G1 by lashing all available fire hoses to the guardrails and setting them to spray. According to the commander's report, the hoses 'were secured with the nozzles pointing upwards and the combined effect when the valves were opened, was most satisfying'.[26] Both of these vessels were commanded by the 22-year-old Lieutenant George Halley, and they did invaluable work during Mosaic. Lieutenant Halley provided a vivid description of G1 as he and his men experienced it from the deck of MWL 251:

> At 1151, although we had our backs turned towards Trimouille Island, we experienced a blinding flash of intense magnitude, followed by a slight burning feeling across the back of the neck and at the back of the knees. This was only momentary and the intensity of the heat resembled the warmth of the sun on a December day. At 'H' plus 5 seconds, the hands were permitted to face the direction of Trimouille Island. On looking around we observed the last stages of the fireball. It resembled a huge oil fuel fire. As soon as it had contracted a thick mass of dark grey cloud rose in a vertical direction at terrific speed. The familiar mushroom cloud soon developed.[27]

Halley noted that as the minutes ticked by, the cloud changed colour from deep grey to light peach. 'The final shape resembled the letter "Zulu". At this stage, the vapour trails of the "Canberras" could be seen as they flew through the radio active clouds.'[28] For the remainder of May 1956, the two lighters operated a ferry service between the mainland and the islands, racking up over 2500 kilometres of steaming.[29]

The AWTSC established mechanisms for monitoring fallout from Operation Mosaic. These were primarily in the form of two different kinds of sampling devices – air samplers and sticky paper samplers. The air samplers used filter paper to capture dust particles from the air, while the sticky paper samplers caught falling particles on their adhesive surface.[30] The air samplers were housed in shelters to prevent overheating or rain damage, and the sticky paper samplers were placed on top of the shelters. The samples were then sent to the Commonwealth X-Ray Laboratory in Melbourne for analysis, using equipment supplied by AWRE.[31] For Operation Mosaic, these sampling stations spanned the country, from Onslow, Broome and Port Hedland on the Western Australian coast, to Alice Springs and Darwin in the Northern

Territory, to Townsville on the North Queensland coast, as well as Sydney, Melbourne, Brisbane, Adelaide, and many other sites, a total of 28.[32]

In the aftermath of G1, those present reported seeing how the blast had devastated the local sea life. An electrician on board HMS *Alert*, RG Maughan, wrote a letter to his mother on 17 May from the islands, saying 'we were not allowed to fish or make freshwater, owing to the water in the lagoon being radio-active. Thousands of dead fish were floating on the water as the tide turned'.[33] This crew member also witnessed G2, which he described to his mother in a letter as 'a real monster' that he estimated to be four or five times bigger than G1, commenting 'it will never be seen on the newsreel'.[34] Despite the prohibition on fishing, one participant, DA Keys, said that it occurred anyway, and that personnel wandered at liberty on the islands between tests, 'although the men did use common sense in this matter'.[35] Keys also said that no protective clothing had been issued, 'except in the case of those entering into the proximity of ground zero. Most personnel were wearing shorts only during the test, and stood with their bare backs to the explosion'.[36] While he was ashore for G1, when he asked his fellow crew members about their experiences, they told him the moment of detonation felt like 'a blow-lamp drawn quickly across their backs'.[37]

Safety was generally lax and equipment that should have been routinely used for atomic weapons testing was not. John Berry, who was called up for UK national service in May 1955, having been working as a scientific assistant at Fighter Command headquarters at RAF Bentley Priory near London, told the Royal Commission about the inexplicable lack of suitable equipment. He had been at Monte Bello from early March until late July 1956, aboard HMS *Narvik*. 'The only radiological measuring equipment I saw being used was the film badge, one of which I was given and which I handed in when I left. Although I was with the AWRE staff the whole time I was there I did not see any use of geiger counters.'[38] When interviewed by the Royal Commission in London, he said that he had not been given any instructions on how to use the film badge.[39] He also asserted that safety equipment was jettisoned on the journey back to the UK – 'I know that on our way home a whole collection of batteries for geiger counters and protective boots were thrown overboard as we were heading back towards Suez because apparently we needed more room in the cold store'.[40] When queried, he did concede that the jettisoning of the Geiger counter batteries could indicate that the equipment was on board but he did not personally see it being used.[41] Like RG Maughan, Berry testified that the cloud from Mosaic G2 was five to six times the size of that of G1.

Journalists were only present for G1. Unlike Hurricane, when members of the press were barred and their attempts to reach Monte Bello unofficially were rebuffed, journalists were permitted to join the official party for the first Mosaic bomb. They were not made to feel welcome, though, either by British or Australian military authorities. Their activities were strictly policed, according to the edicts of the UK Ministry of Supply. Instructions provided to the Monte Bello Working Party indicated that only limited information was to be provided, and the fact that there would be two explosions had to be concealed, even though stories had already been run in the British press suggesting that there would be two. The minutes of the ninth meeting of the Monte Bello Working Party assert that 'the press must not be present at any explosion after the first and it must therefore not be disclosed that there is the possibility of a second'.[42] At an earlier MOSEX meeting, the intention to deceive the press was laid out. Should G1 fail to detonate, 'it could be explained locally that a last minute decision to delay the firing had been taken'.[43] If the blast had a lower yield than expected, the authorities would nevertheless provide a signal that indicated that the test had been 'fully successful'.[44]

This suspicion towards the press came from the top and influenced all involved with the tests. A typical response to them is found in the Report of Proceedings by the commander of HMAS *Fremantle,* Lieutenant Commander Julian Cavanaugh, and an associated handwritten note attached to this report, addressed to Hugh Martell. Fourteen 'gentlemen of the press' were onboard *Fremantle* to witness G1.[45] The contempt in the note written a few days after the report is unmistakable: 'The Press are never satisfied with a handout about something of which they know little or nothing. If their intelligence was sharp enough to be affronted by it, it should also have made them realise the problem of security'.[46] Also on board *Fremantle* with the loathed media was an assortment of Australian military VIPs, including Vice Admiral Roy Dowling, Captain Jack Mesley, Lieutenant General Sir Henry Wells, Group Captain Brian Eaton, Lieutenant General Alwyn Ragnar

Garrett and Air Vice Marshal Ellis Wackett, eager to see the new weapons of war in operation though rather surplus to requirements themselves.[47] VIPs were tolerated but generally seen as 'tourists', nosey types who were not essential to the operation. During the journey from Onslow jetty out to the islands on Tuesday 15 May, Commander Cavanaugh held a press conference to brief the assembled media on the coming test. When the ship reached Monte Bello, the military VIPs were transferred to HMAS *Junee*.[48] The journalists remained onboard *Fremantle*. Once the device had been detonated, the journalists got to work. When 'the rattle of typewriters had ceased' just after 1 pm that day, their stories were despatched to MRL 252, then to *Narvik* for censorship,[49] then onward to the mainland.[50] Cavanaugh did not enjoy hosting the journalists on his ship:

> The enquiring minds of the Press made my task a difficult one, particularly at the press conference. I made no statements other than those authorised by instruction of the Commodore, Special Squadron [Martell]. However, it was noticeable that a proportion of classified material had been readily and easily amassed by members of the press, to which no repudiation or confirmation other than 'no comment' retort could be given. This however did not necessarily satisfy their natural sagacity and in effect classified material could have been published purely on their assumptions, some of which may have been quite correct.[51]

Although they had been told that they had to limit themselves to an impossible 40 words per correspondent, owing to restrictions on telecommunications from Monte Bello, few bothered with this imposition. As soon as G1 went up, the journalists didn't hold back with their copy. Martell complained that the reports 'were very lengthy and no attempt had been made to limit them to the 40 words specified. Some despatches consisted of several pages of foolscap'.[52] The captain of HMAS *Junee*, Lieutenant Commander Smith, caught hell from Martell for prematurely releasing press materials upon the detonation of the first Mosaic bomb. Smith had overlooked a paragraph in the trial orders that had given an exact timeline for the dissemination of press material.[53] Martell wrote a stiff note to him, saying '[c]are must be taken always in reading operation orders thoroughly, and in interpreting them with care and in the manner intended. In this case there can be no excuse for misreading or misunderstanding' the relevant parts of the Mosaic trial orders.[54]

One of the challenges facing commanders of the ships present for lengthy periods at Monte Bello was morale. A bay on North Delta Island (now Campbell Island), to the south-west of where G1 was to be fired, was netted to stop sharks and stonefish getting in, and a beer canteen was set up on the beach.[55] This crescent of sand was called the Lido, arising from the sort of tongue-in-cheek joshing that often came into play with British nicknames. The original Lido was a glamorous beach in Venice, but this beach – despite its stark natural beauty – had nothing in common with its Venetian counterpart. At the Monte Bello Lido, beer was sold 'in unlimited quantities' and a large marquee was set up for the troops.[56] To observe the proprieties, chief petty officers and petty officers were accommodated in their own 'smaller and slightly more secluded shelter' a bit away from the men.[57]

Despite the best of intentions, though, this attempt at providing popular recreation was something of a flop. The bay was a long boat ride from where the ships were anchored, the tides meant that good swimming was only available for about two hours, and in any case bad weather soon tore out the netting.[58] The Lido became overwhelmed by seaweed, causing its popularity to wane.[59] An attempt at establishing regular cricket and football matches on a pitch built on Hermite Island, south of Alpha, was likewise unsuccessful due to a general lack of enthusiasm, probably also related to distance from the ships.[60] Trimouille Island, only a few hundred metres away, was still radioactive from Operation Hurricane, so landing there for recreational purposes was not permitted.[61] Nevertheless, and in common with military forces throughout history, the men found ways to amuse themselves, whether being pulled along behind a whaler (a small service boat used for moving quickly around the islands) or fishing in the lagoon before the first blast (and surreptitiously after the blasts). On board the ships, gyms were set up and the crew could play deck tennis or have boxing matches. There was even .22 shooting on the forecastle or skeet shooting from the gundeck: 'the latter is becoming extremely popular'.[62] The sailors did not think much of the little town of Onslow, through which pretty much everyone passed on the way out to the islands, and the RAAF camp was based there. The commander of HMS *Alert* was dismissive, saying that it 'offers nothing apart from beer at four times the price at which it is sold onboard'.[63] The same jaundiced eye was cast over the major role of HMS *Alert* at Mosaic, housing many of the operation's scientists, as well as the Australian safety committee. As the commander noted in his report, considering the 'long and dreary period of waiting and the ship's uninspiring role in the operation – that of a floating hotel for the boffins – the morale of the ship's company has been remarkably high throughout'.[64] Normally, HMS *Alert* was the yacht for the commander-in-chief of the Far East Station for Britain, not a boarding house for scientists, a distinct come-down.[65]

Only three Australians were aboard the lead ship, *Narvik*, for the Mosaic trial: the meteorologists Bob Southern and Henry Phillpot, and ASIO security officer Leo Carter. Both Southern and Phillpot would have long careers with the Bureau of Meteorology, Southern as regional director in both the Northern Territory and Western Australia and Phillpot leading the International Antarctic Analysis/Meteorological Research Centres from 1959 to 1969.[66] On board *Narvik* the two Australian meteorologists monitored for possible tropical cyclones and gave twice-daily briefings to Martell, Southern joining the ship before Phillpot.[67] Southern's departure for the islands, along with everyone else on *Narvik*, was delayed by Tropical Cyclone Adam, which tore through Onslow and the Monte Bellos before heading down the Western Australian coast, doing considerable local damage as it went. Southern viewed *Narvik* with some scepticism when he saw the vessel for the first time, as the cyclone was still menacing the coast and he knew he would be boarding soon. The vessel's foredeck was level with the Fremantle wharf, making it look small and insignificant, especially given the presence of a destructive storm just up the coast. However, once out at the islands the ship looked 'considerably more impressive, in comparison with the low-lying Montebello Islands, when we anchored in the lagoon on the evening of Thursday, 8 March'.[68] He began his six-hourly weather observations the next day, along with hourly wind speed and direction logs. Although he was an Australian, even Southern had difficulty with the heat, but the British crew found it unbearable. Southern noted that temperatures in the office where he worked topped 41 degrees over ten successive days, rising as high as 43 degrees.[69] For the visit of First Sea Lord Louis Mountbatten, Southern wore an emblem on his hat depicting a weather balloon and a bolt of lightning, designed by petty officer Ron Cumper.[70] The emblem attracted the attention of Mountbatten, who quipped, 'Well, I can at least tell what job you're doing'.[71]

The lead-up to G2 was agonising. The weather was uncooperative, and the task force was already operating in marginal climatic conditions anyway. The Monte Bellos were simply not suitable for a mid-year detonation, a fact that had been known by the British authorities since before Hurricane. Given time pressures created by the forthcoming Maralinga tests and, even more weighty, Operation Grapple in the Pacific where *Narvik* was needed, the weather-related tension there grew with each passing day. At one point there was 'a brief mutiny' on board *Narvik*, as later reported in the British media, when deckhands refused to obey an order to move a landing craft.[72] Everyone involved was restive. The added complexity of having the AWTSC making their presence felt, with that menacing veto they could wield on firing, made for difficult days out at the islands, particularly for the leaders of the operation.[73]

One source of pressure was absent, though. For G2, the word came down from MOSEX that there would be no VIPs or press in the official party, to the relief of those who had had to deal with both during G1. The first G2 rehearsal, under the codename Glimmer Delta, was held on 28 May. The fissile material for G2 arrived at Onslow on 7 June, and another secret operation was undertaken to get it to the islands. This time it was transferred to a hut on Alpha Island to await the other components before assembly. A minor panic arose when the hut proved to be leaky, and tarpaulins were called for as tropical rain poured down.[74] The downpour flooded the floor of the hut and delayed assembly of the device by 12 hours.[75] The rain cleared sufficiently and assembly of the weapon was complete on 10 June ready for it to be hoisted up the 31-metre tower on the north-eastern end of Alpha Island. However, the wind continued to be unfavourable and it was not until 13 June that the meteorologists had any good news to share of a possible break in the weather in coming days. That optimism was somewhat misplaced, and several days of intense frustration ensued as bad weather kept rolling in. Finally, overnight on 18 June the weather suddenly improved and stand-by was called. The firing order code word for G2, Flashlight, was transmitted and the countdown began at 10.10 am. Four minutes later, G2 exploded. Back in the UK, Whitehall received the code word just two minutes after detonation indicating that the blast was successful.[76] Scientific personnel and others on Hermite Island were quickly evacuated directly after the firing. They were first placed on an LCM, then, using an evacuation route decided in advance, the scientists rendezvoused with MWL 251, *Gunga Din*. All ships proceeded south to their anchorage in the lee of North Sandy Island.[77] Martell described the scene among the scientists and military personnel as the operation drew to a close:

> A tired and dirty, but jubilant, Hermite party crashed thankfully on to their bunks and the Sapper tower party regaled their chums on the messdecks with stories of the last moments of Alpha tower. The main topic of conversation unquestionably centred around the journey home which had now become a reality instead of a myth.[79]

Although no journalists were permitted in the official party for G2, they were still observing events as best they could from the mainland, although not this time from the 1952 lookout. One (unnamed) member of the press, who reported on Mosaic G2 from Onslow, provided a description for the ages.

> The cloud came up like a giant ice cream cone. At 10.17 (local time) the cloud was like a double mushroom. Then it was like a woman with a turned-up nose and bobbed hair, still attached to the ground by an elongated neck. By 10.18 the top had formed into a cauliflower like a lone ... thunderhead above the horizon. When the cloud first appeared it was darkish orange-grey and to watchers on Onslow Jetty it had none of the colours shown by the previous cloud.[79]

In the mid-1980s, counsel assisting the Royal Commission, Peter McClellan, would challenge William Penney about the foolhardiness of Operation Mosaic, saying 'taking the standpoint of looking at what happened after the event, that both the Mosaic explosions ... were fired in conditions which ultimately showed they did not fulfil the safety expectations. You were aware of that?' Lord Penney: 'I know it was not perfect, yes'.[80] That lack of perfection had, for a time, plunged the Australian population into fear.

One source of considerable disquiet were the actions of a man with a Geiger counter. At the Royal Commission, miner Stewart Stubbs gave evidence of what happened when he was working at his mine near Marble Bar in Western Australia and detected the Mosaic fallout. Stubbs was aware of the tests at Monte Bello, and at the time of G2 claimed to have heard a couple of explosions.[81] Marble Bar is roughly 400 kilometres east of the Monte Bello Islands. Whether he actually heard the sound of Mosaic G2 is not clear. He also noticed the arrival of a low cloud. Concerned about the possibility of radioactivity, he got out equipment that he used for uranium prospecting on behalf of the Bureau of Mineral Resources. In drizzling rain, Stubbs used his two Geiger counters to measure the radioactivity on the ground and on the top of his vehicle. Both instruments went 'off the scale'.[82] Stubbs was an experienced miner and prospector, and he said that the only other time he has seen Geiger counter readings go that high was when he had used them on a piece of the radioactive, uranium-rich ore pitchblende from the mine at Rum Jungle, in the Northern Territory.[83] Stubbs radioed through to the airport at Port Hedland, about 200 kilometres away on the coast. He recalled speaking to a scientist he thought was called 'Penney', although that could not have been William Penney as he was not present for Mosaic. He then said he was put through to someone who was 'in the air in a Canberra bomber', also seemingly unlikely.[84]

> I told him what had happened, that these black clouds had come over and that we had a drizzly rain and that ... there had been very high readings on the Geiger counters ... I said 'Is this dangerous to people up here? What is the position?' He said 'I can't comment.'[85]

While it was improbable that he was put through to a person in an aircraft, he may well have spoken to someone connected with the air operations for Mosaic. Somehow a journalist found out about this interaction, through Stubbs' name apparently being mentioned in a report from the British government to the Australian government. One newspaper quoted the acting prime minister Arthur Fadden as saying:

> 'While it cannot be stated with certainty what caused the increased count on the private resident's geiger counter at Marble Bar, it is possible that some deposit from the test took place in that area. What can be stated without equivocation is that any such deposit is completely harmless and similarly in relation to the report of radioactive rain, there is no cause whatever for concern.'[86]

The Royal Commission later confirmed that fallout from G2 had travelled across the continent. 'The sticky paper samples showed that the high altitude cloud travelled right across northern Australia at a latitude of about 20 degrees S and that after 60 hours it was over the Coral Sea, clear of the Australian coastline.'[87] Onslow received significant fallout, as did Broome. The test was not safe, despite public pronouncements. What the Australian public did not know was that Mosaic G2 was built to be extremely powerful, and its extra-large load of radioactivity was part of the experiment. All the public knew was that it looked awfully like something had gone terribly wrong.

CHAPTER 9

WHAT WAS G2?

> Mr Butement, Sir Leslie Martin and I went and physically inspected [the Mosaic G2 device] just before it was armed. We stood, if you like, and admired it. I do not want that to be misunderstood; but in truth that weapon was a very beautiful piece of highly sophisticated engineering and I think that the only word to describe it is beautiful.[1]
>
> ERNEST TITTERTON, EVIDENCE TO THE ROYAL COMMISSION

> Questioned on an Adelaide University report that atomic rain had fallen in South Australia, very possibly as a result of the Monte Bello test, Liberal Minister for Supply Beale could only reply ... 'I don't believe it.' Maybe he is waiting for a couple of corpses to convince him.[2]
>
> JOHN McISAAC, SYDNEY *TRIBUNE*, 1956

The story of G2 captures the secrecy and deceptive tendencies of the British government, which was testing deadly weapons in another country but did not want to share its plans or its test results with that country. Australia was right to harbour some suspicions about the Mosaic tests. Both G1 and G2 were boosted fission weapons, although the design of G1 meant that it had nowhere near the explosive power of the second test in the series. G1 and G2 were essentially huge 'physics experiments',[3] undertaken as Britain attempted to identify the most fruitful direction for its newly approved megaton weapons program. The technical difficulties facing Penney and his team in creating the world's most powerful weapons involved the extreme temperatures and densities of the material in the weapon, something that William Penney described as a problem 'of immense technical difficulty'.[4] Achieving the temperatures required for fusion – 100 million degrees Celsius – within the confines of a relatively small bomb assembly is significantly challenging. The only way to do it in a standard hydrogen bomb is to use a fission bomb as the trigger. But this was not the design of the Mosaic bombs, because strictly speaking they were not hydrogen weapons.

However, they did have those enigmatic light elements, exactly what would be found in a hydrogen bomb. G1 had a modest explosive yield of 15 kilotons, substantially smaller than Hurricane's 25 kilotons, and with little of its explosive power coming from the addition of light elements to the fission assembly.[5] The second bomb was another matter. Mosaic G2 was different to any other weapon tested by the British in Australia. G2 contained uranium as a 'tamper', while G1 had a lead tamper.[6] A tamper is a dense layer of material wrapped around the fissile core, used in some nuclear weapons designs to control and prolong the chain reaction, delaying the disintegration of the core.[7] By using only a lead tamper in G1, the bomb makers would be able to see what effect, if any, the light elements had on the strength of the fission detonation. Would these fusion materials still significantly enhance the explosion, even with an inert tamper? This had to be tried first with lead, to isolate the effects. However, adding a uranium tamper to G2 was thought to have the potential to greatly increase the explosive power of the weapon. Unlike lead, natural uranium-238 can undergo fast fission under bombardment by high-energy neutrons, thereby adding substantially to the explosive yield. This isotope of uranium is not fissile and therefore cannot sustain a nuclear chain reaction, but it is fissionable under such conditions.[8] This was a recipe for a high-power weapon.

The drama and controversies around G2 all date back to the momentous decision two years earlier to travel a new nuclear path. In April 1954, the British government called for detailed information on the H-bomb, including insights into whether the UK would benefit from developing its own version. A secret Ministry of Defence paper set out the case, and this paper was discussed at a Ministry meeting of the Working Party on Operational Use of Atomic Weapons chaired by Sir Frederick Brundrett and attended by William Penney and other officials. The assumption in the paper was that, should a nuclear war break out, Britain would be operating jointly with the US, but would primarily be taking care of its own neighbourhood; 'our contribution would probably be to deal with targets of general importance to ourselves, leaving the general strategic offensive to the United States'.[9] Despite its differences with the US, and the obstacles put in the way of joint nuclear development, Britain would still work closely with America in the event of an outbreak, and planned to contribute its own weaponry to the effort. This required the best technology. Penney put forward his view, in light of limited British resources, that it would be possible to make a 10-megaton weapon from fissile material equivalent to what was needed to make three Blue Danube bombs or two 'mixed' (boosted fission) bombs. However, the design of a British megaton weapon would require uranium-235, a natural – and rare – fissile isotope that would not be available in sufficient quantities until mid-1956. This meant that a megaton test could potentially be carried out at the earliest in spring of 1957.[10] Until then, the Blue Danube fission weapons would continue to be produced, said Penney, and in fact the existing stock of the original weapon could be repurposed once the new weapons were being produced, thus causing 'no interruptions in our capacity to drop atomic weapons'.[11]

In September 1954, a few months after the political green light had been given to the British H-bomb, Brundrett's Working Party on Operational Use of Atomic Weapons met to progress this political imperative. This meeting assessed what was needed in addition to the development work already undertaken. Brundrett told the working party in a memorandum, in response to discussion about the power required of the British H-bomb, that the country's aim was to be able to destroy the enemy's centre of government or vital industry.[12] Brundrett asserted that a 5-megaton weapon would be sufficient for this task.

The British authorities referred to their primary fission weapon, Blue Danube, as Mark 1. William Penney reflected in a top-secret document attached to the minutes of the September meeting about where they were in their nuclear armaments, two years on from the first test at Monte Bello but considerably modified since then. 'Most advanced in our programme is the Mark 1 bomb, now in small quantities with the R.A.F. Several improvements will be introduced over the next few years and this weapon, with its improvements, will be the only one available to the Services until the end of 1957.'[13] This document discussed the next evolutionary development in British armaments, Red Beard (which had originally been called Javelin). This tactical nuclear weapon, much more compact than Blue Danube, was to be the triggering device for the new British H-bomb. Red Beard was one-fifth the size of Blue Danube but with the same powerful output.[14] This weapon's new implosion system, said Penney, was a step up from the original design, and indeed the engineering breakthrough that had led to Red Beard could also be applied to the old Blue Danube design.[15]

The ideas around the use of boosted fission had their origins in scientific discussions at AWRE. In the lead-up to Mosaic, Penney and co were working on two design types, known simply as Type A and Type B. Type A was a boosted fission weapon and Type B was a true hydrogen weapon that contained a fission trigger as part of its two-stage detonation, given the name Teller-Ulam after its designers, Edward Teller and Stanislaw Ulam. Type B devices were 'thermonuclear', a designation that arises from the need to create extraordinary temperatures inside the assembly. On the other hand, a Type A boosted fission device was not thermonuclear but was a kind of hybrid. Penney and other British authorities became interested in seeing if Type A might work because it was technically less complicated than a true hydrogen bomb. The two bomb types were related but with significant differences. As historians Lorna Arnold and Mark Smith put it, 'The distinction between boosted fission weapons and thermonuclear weapons is a fine one, but should not be overlooked. The linking theme is the use of light elements to produce large explosions'.[16] This distinction is crucial to understanding the nature of G2, and in discounting later claims that G2 was a thermonuclear device.

In this document, Penney revealed that he and his AWRE team were working on the 'boosting principle', in which small quantities of an isotope of lithium (lithium-6) could be added to a fission weapon to make the bomb much more powerful. Penney stated that progress was 'reasonably good but there is a lot of work to be done'.[17] Here were the first hints of the nature of Mosaic G2. Penney needed to be sure that Type A was indeed not a viable option for Britain, and that the British megaton weapon would be of Type B design, although it might stand as an interim type depending on how the experiments went. According to historian John Walker, 'it was vital to be sure that UK calculations on boosting the yield of a fission weapon using light elements would not in fact lead to large yields'.[18] Although Mosaic confirmed British calculations that boosted fission was not a direction they wished to take, 'the two tests did make essential contributions to AWRE's knowledge of fusion reactions and its results were crucial for planning [Operation] Grapple'.[19]

Notes from a UK Chiefs of Staff Committee meeting in 1954 state that while they were 'reasonably confident that we can make such a weapon [Type A, boosted fission], we much prefer to go for the second type, B, because it gives a much bigger yield for expenditure of fissile material and money. However, the first type does have two advantages, comparative simplicity and comparatively smaller demands on quantities of fusion material'.[20] At this point, British expertise was not yet at Type B level, according to Penney. A way of advancing their knowledge while potentially eliminating a particular design style from their range of possibilities was to try Type A first. The exercise would also help Britain to talk up its capabilities to the country it wanted most to impress, the US, its importance being, according to Lesley Wright and others, its ability to 'enable politicians to exaggerate the technical capability of the British armed forces, hence increasing perceptions of British capability and her prestige'.[21] In the end, they did go with Type B, having attempted their version of Type A as part of Mosaic.

A fortuitous back channel to the Americans on H-bomb development had opened up, just as the British were wrestling with multiple technical challenges. In 1955, Penney had led a British delegation to the Tripartite Conference on Atomic Weapons Effects in Washington, involving delegates from the US, Canada and the UK. He had returned with bundles of documents relevant to his work, including some restricted data. Penney said that the information 'had only been made available by the U.S. as a great concession, considering the present U.S. legislation [the McMahon Act]. The fact that we had the information should not therefore be disclosed even to U.S. authorities who had not taken part in the conference'.[22] The British delegation had been allowed to take restricted data home with them with the tacit consent of US participants, so long as that information was not made known more widely. American nuclear weapons scientists seemed to be finding low-profile ways to provide Britain with information while getting around the McMahon Act. Penney's Manhattan Project–era colleagues were no doubt helpful in this respect. An obituary published by the Royal Society after Penney's death told of the importance of his Manhattan Project years in being able to run the British weapons program: '[Penney] kept in close touch with his American friends and was able to sit up drinking with them half the night. There is little doubt that he and he alone gleaned in this way much information about the American programme'.[23]

By February 1955, many aspects of the H-bomb program were coming into better focus and its objectives were clearer. Penney wrote a top-secret note for the controller of atomic energy, Sir Frederick Morgan, to clarify where he was with the development program and to ask Morgan if he saw the need for any further refinements. While Type B was favoured, with a 5-megaton weapon on the drawing board, such a weapon would be 'unavoidably heavy' according to Penney.[24] In addition, it was going to use large quantities of lithium-6, which had proven to be unexpectedly difficult to manufacture. The H-bomb program would need several tons per year, a difficult and expensive undertaking, although rather less was required for a Type A bomb.[25] Accordingly, Penney pushed for a test of the boosted fission weapon, based on additional work at AWRE: 'a further study has been made of Type A, with the result that very strong arguments can be advanced for the development of Type A bombs before Type B'.[26] These weapons would be much lighter, needed far less lithium-6 and could potentially produce a yield of 1 megaton. Should testing go well, Penney predicted that it would be possible to produce a deployable 'free-falling' megaton weapon for the RAF by 1959.[27] By comparison, a Type B bomb was unlikely to be deployed before 1961. (In the brief period that an interim high-yield but non-boosted fission weapon was offered for deployment to the RAF – which didn't actually want it – it had the codename Violet Club.)[28] Penney made the point that 'the development of Type A before Type B is the natural line of approach, and is undoubtedly one adopted by the U.S.A. and probably by Russia as well. To this extent, the approach through Type A to Type B is technically sound'.[29] At this point, Penney was clearly advocating to at least try the boosted fission design before committing to the more expensive Type B, which would take two more years to be able to place in the British arsenal.

Type A would not give Britain full expertise in hydrogen weapon production, but would be experimentally useful as well as providing an option for a stop-gap weapon. But they really wanted the full H-bomb. The ultimate aim was to develop a ballistic missile to carry thermonuclear warheads, and this too would require a lighter weapon. The V bombers were expected to become obsolete early in the 1960s, thus removing the RAF's only delivery system for a full-scale H-bomb. Penney spelled out the potential advantages of a boosted fission weapon: production would start earlier, 1-megaton weapons would be available in slightly larger numbers, the delivery system was more flexible, and delivery of the smaller bomb would be much safer for aircrew. Most of all, Penney said, the UK 'would demonstrate to the world the possession of a thermonuclear bomb much earlier than otherwise and this may be politically of very considerable importance'.[30] His foundational Blue Danube weapon could be refined as the basis for the Mosaic bombs. He also recommended that work proceed on Red Beard, both as a tactical bomb in its own right and also as a 20-kiloton trigger for the Type B thermonuclear bomb, but that initial priority in the thermonuclear field be given to a 1-megaton warhead based on Type A design.[31] The course towards Mosaic was set. The British were now focused on weapons of frightening yield, and Mosaic G2 would be their biggest yet.

The G2 detonation was the culmination of a process that effectively eliminated the boosted fission design, clearing the way for a full-scale British hydrogen bomb test in the Pacific the following year. Some estimates put the yield of this weapon as high as 98 kilotons, about six times the explosive strength of the bomb dropped on Hiroshima in 1945, although 60 kilotons remains the official figure. The methods for measuring yield in 1956 had some problems associated with them, and hence some of the confusion. Confusion has also been caused by the ongoing, inexplicable refusal by the British government to release all relevant documents, most likely for political rather than proliferation reasons. The yield officially accepted by the Royal Commission was 60 kilotons, after some brief and arguably inconclusive discussion during questioning. However, what is clear from the documents is that Britain was expecting a larger yield than that. During the Mosaic planning stage Penney discussed with Sir Edwin Plowden, chairman of the United Kingdom Atomic Energy Authority, that 'we do not know exactly what the yield is going to be because the assembly is very different from anything we have tried before'.[32] However, and while acknowledging it could be less, he disclosed that AWRE was planning for a yield of around 80 kilotons, 'which is the safe upper limit'.[33] The enormous explosive strength of this weapon and the ongoing secrecy surrounding it mean that G2 remains one of the enduring mysteries of the British test series in Australia.

The issue of the G2 yield has provoked more controversy than any other British bomb in Australia, not helped by the insertion of misinformation into the narrative. In 1984, *New Scientist* magazine published a story saying that G2 was larger than had been disclosed to the Australian government, noting Ministry of Defence figures that indicated a yield of 60 kilotons. However, some of the assumptions in the story may be flawed. The article appears to have been written without knowledge that the Australian government had been told in 1955 that G2 would not exceed two and a half times the yield of Hurricane, so it was no revelation to say that the explosion was 60 kilotons.[34] At the time of Mosaic, members of the AWTSC – after some pressure – were privy to the yields of British weapons and therefore knew what 2.5 times Hurricane meant. The story, by Steve Connor, cited a letter written by Thatcher government Minister of State for Defence Procurement Geoffrey Pattie to Labour peer Lord Brockway that he claimed revealed the test to have a yield of 60 kilotons, and that this was three times the size that the Australian government had been advised. The *Canberra Times* picked up on this story two days later, having obtained additional Australian government comment. The minister for resources and energy, Peter Walsh, was already under pressure over whistleblowing and investigative journalism around the aftermath of the British tests in Australia, and would soon reluctantly call a Royal Commission to examine exactly what had been done on Australian territory. He was quoted in the *Canberra Times* article as saying, 'I have instructed my Department to press the British Government for further information in light of these reports and for further information to be made publicly available'.[35] Neither the *New Scientist* nor *Canberra Times* stories mention a claim of 98 kilotons for G2, only (erroneously) that the Australian authorities expected it to be 20 kilotons when it was 60. The stories do, however, neatly point to the fact that the Australian government in the 1980s did not have detailed knowledge of what the British actually did during their Australian tests in the 1950s, including what the earlier generation of Australian politicians and the Safety Committee were told by the British authorities. The most authoritative sources for figures higher than 60 kilotons are the technical calculations made by the AWRE itself.

Secret planning documents prepared at the AWRE reveal that the British bomb makers fully expected Mosaic G2 to be well above 60 kilotons. Even though they assured the Australian government that the bomb would be no more than 62.5 kilotons, they were actively planning for it to be much bigger, without ever informing Australia. A joint report from the Atomic Energy Authority and the AWRE suggested that the Mosaic 'explosions are expected to have upper limits of yield equivalent to one and four nominal in that order'.[36] The term 'nominal' in nuclear weapons lingo refers to 20 kilotons, so the authorities were planning for an 80-kiloton weapon for the second Mosaic test, something Penney confirmed in his letter to Plowden. This document devised safe distances from the explosion based on this expected large yield. Using these calculations, the report suggested 'that flying be suspended within a distance of 400 miles for 10 hours and that aircraft between 200–600 miles fly below 5,000 ft. for 24 hours after burst'.[37]

Three days after the firing of G2, the scientific director, Charles Adams, sent a top-secret message from *Narvik* to William Penney back at Aldermaston, providing preliminary yield measurements. The message said that the 'radius time curve gives 120 kilotons using free air formula'.[38] Adams was referring to a technical method for measuring yield that relied on the rate of increase of the fireball, one that had already successfully measured the G1 yield. Adams continued, 'Quality excellent, observation inaccuracy at most 10%. In view of G.1. result consider significance must await analysis of scale effect of tower height. Development other records therefore deferred until arrival U.K.'.[39] He also said that measurements taken at H1 and aboard *Narvik*, respectively, gave 90 kilotons and 107 kilotons. Hinting at an inhouse competition to guess the yield, Adams suggested that a 'sweepstakes figure' would be 90 kilotons.[40] The ambiguity surrounding the G2 yield was off and running.

A secret document prepared for William Penney by AWRE scientist Roy Pilgrim in February 1957 reviewed the work done to date on analysing the Mosaic yields. Pilgrim noted that various kinds of measurement systems were used, raising the issue of the relative reliability of the kinds employed, although he noted that the figures they were working with were giving good results for Mosaic.[41] In this document, Pilgrim gives two values for G2: 100 kilotons (with the note that this was '[n]ot yet accurately assessed') and 62.5 kilotons, plus or minus 1.2 kilotons.[42] No explanation is given in this document for the differing figures. He made the point that he had not had the opportunity to take into account the 'various adjustments necessary in determining the total yield from the Radio Chemistry data', and so he had not included this in his calculations. That radio chemistry data is still not available in the public sphere, so what that data did to the final calculation is not known. Adding to the confusion, a secret Admiralty report of March 1957 gives the G2 yield as 60 kilotons,[43] based on data supplied by AWRE.

One of the reasons confusion remains about the G2 yield is because yield measurement techniques in the 1950s were less effective and accurate than they later became. In fact, the sheer ferocity of the G2 blast damaged some of the measuring equipment, rendering it unreadable. Nevertheless, values from two of the methods were obtained and published in a secret AWRE report in June 1957, months after the Admiralty report that gave the yield as 60 kilotons. The report was authored by AWRE scientists R Potter and AC Purdie. They employed two methods: measurement of pressure at distances from the detonation using tubes and gauges, often known as the 'toothpaste tube' method; and measurement of the time taken for the blast pulse to reach various distances from the detonation. The first method estimated that Mosaic G2 yielded 98 kilotons, while the second had a more conservative estimate of around 80 kilotons (still well above the 60-kiloton figure accepted at the Royal Commission). The report does not mention 60 kilotons as a possible figure; the lowest yield figure derived from the data is 74 kilotons. The second method was particularly unreliable, as Potter and Purdie noted, saying 'the accuracy of the method is ... somewhat limited and magnitudes obtained by it should be regarded as purely indicative of weapon size'.[44]

The methods described seem remarkably basic. The toothpaste tube method, which had also been employed at Emu Field, used aluminium tubes just under 9 centimetres long and 1 centimetre wide. They were the same kinds of tubes supplied to toothpaste manufacturers at the time.[45] The tubes were arranged in sets of nine in baffles about 30 centimetres long and 12 centimetres high. The tubes collapsed from the blast wave and were later plunged into water to measure displacement and thus determine their volume.[46] This first method also employed a thin duralumin diaphragm 6 centimetres in diameter. Duralumin is a lightweight alloy of aluminium then used in aircraft construction. The centre of the diaphragm was connected to a stylus that was able to write on a strip of moving celluloid film.[47] These items were placed on Trimouille and Alpha islands for Mosaic G1 and G2, respectively, at various distances from ground zero, the closest being 457 metres and the furthest 2072 metres. Efforts were made to place the devices on flat areas to avoid the interference that could be caused by hills.

The second method, which measured transit time, recorded three separate events at a point 16 kilometres from the blast: the flash upon detonation of the weapon, a small acoustic pulse from a subsidiary explosion (used to ensure an accurate measurement of the velocity of sound) and the weapon's blast wave.[48] The subsidiary charges were placed on Beaver Hill, an interior rise on the western/central part of Trimouille Island, for G1, and North Delta Island for G2. The Beaver Hill charge was tested several times before G1 to check accuracy. The measurements were recorded on electrically conductive paper known as teledeltos, and were taken both on Hermite Island and aboard HMS *Narvik*.

For G1, a much smaller blast than G2, the methods appeared to function adequately, although the second method suffered from some technical problems that probably ruled out the values it obtained. The tubes and diaphragms gave a yield reading between 13.8 and 19 kilotons, while the transit method gave a high reading of 29 kilotons. The accepted strength of Mosaic G1 is 15 kilotons, so the first method appears the most accurate.

However, the methods suffered from more technical issues during the detonation of G2. A number of collapsable tubes were damaged by flying rocks, so only a few sites provided usable data. All but one of the diaphragms failed as well. Nevertheless, yield figures were obtained, in the range of 94 to 102 kilotons, allowing for a margin of error on either side.[49]

For G2, the transit time apparatus functioned correctly and provided good quality records. When the data were analysed, the AWRE scientists came up with a yield for G2 of between 74 and 82 kilotons using this method. With a bit more number crunching in which both methods were given equal weight, the scientists settled on final yield figures of 15 kilotons for G1 and 98 kilotons for G2. A note in the report indicated that subsequent use of the same two methods at Operation Buffalo at Maralinga later that year 'indicated that it is reasonable to attach equal weight to the results given by them'.[50] So why did the Royal Commission accept the 60-kiloton figure, thus confounding scholars and others into the future? Initial correspondence from Anthony Eden to Robert Menzies had assured the Australian prime minister of an upper limit of 2.5 times Hurricane yield, to a maximum of 62.5 kilotons, and it is possible that Penney and others did not wish to contradict that official advice. Penney was questioned about the Potter Purdie report during the Royal Commission, and dismissed the toothpaste tube data as inaccurate.[51] He did not comment on the transit method, which returned a range of between 74 and 82 kilotons, essentially his own projection before the test was held. Interestingly, Penney made the point several times in his testimony that he was not present for Mosaic, indicating his distance from it since Charles Adams was running the show. In his testimony, he often seemed uncomfortable talking about Mosaic at all, even though as head of the AWRE he was ultimately responsible for its conduct. The 60-kiloton figure was offered to him by counsel for the Royal Commission during his testimony, and while he did not champion it passionately, he allowed it to stand.

In this way, 60 kilotons was established as the accepted figure. The 98-kiloton figure clearly arose from the Potter Purdie document, which was a technical report based on objective measurements made during Mosaic and which have not, at least in the available public record, been contradicted by any other data from the AWRE. It is entirely possible, however, that the transit time measurements were indeed the most accurate, and accorded with Penney's own prediction of a yield of around 80 kilotons. An official history written by Lorna Arnold and Mark Smith suggests that the AWTSC was advised before G2 of the 80-kiloton estimate (with an upper limit of 100 kilotons).[52] AWRE scientist Edmund Drake Seager, who had a scientific role at various British tests but not Mosaic, was approached by the Department of Defence in 1985 to comment on media reports about the 98-kiloton figure at the time Penney was questioned by the Royal Commission. He maintained that the two measurement types undertaken by Potter and Purdie were not reliable, and that more reliable figures were those based on the rate of fireball growth and radio-chemical yield.[53] Exactly what data those two other measurement techniques revealed has not been publicly disclosed, although that early figure of 120 kilotons sent to Penney by Adams in the days after G2 is based on measurement of fireball growth. Until all data are released, the speculation must remain, with the caveat that the Potter Purdie figures are on the official record, without contradiction apart from negative comments by Penney at the Royal Commission about the accuracy of the toothpaste tubes, backed by Drake Seager. Relevant files at the National Archives of the UK are listed as being retained by the Ministry of Defence. The Potter Purdie report is also not currently available publicly at the British archives, while being freely available at the National Archives of Australia.

Even if 98 kilotons is accepted as the true yield, this was nowhere near the strength of a true hydrogen weapon. Those weapons are measured in the megatons, one thousand times a kiloton. The largest ever hydrogen weapon tested, the Soviet Union's Tsar Bomba in 1961, had a reported yield of 50 megatons,[54] hundreds of times larger than the highest G2 value. Despite some speculation that the British did actually test a hydrogen bomb in Australia in contravention of the country's agreement with Australia, the reality is more nuanced. Instead, and possibly in contravention of the *spirit* of the British agreement with the Australian government, G2 was a halfway house, a weapon that was created to investigate whether H-bomb power could be attained without having to create a genuine H-bomb. In the event, all G2 did was convince the British bomb makers that they did in fact need to make a full-strength H-bomb.

Some effort was expended by the British in assuring the Australians that there was no cause for concern about this abnormally large test. A telegram from the Commonwealth Relations Office to the UK High Commission stated that the

> Australians can be reassured. A hydrogen bomb in normal parlance is a weapon of large yield, that is in the megaton range, which employs the fusion reaction of light elements on a large scale. The proposed tests are not of this character but are of fission weapons used as vehicles for certain diagnostic and experimental tests.[55]

Despite being happy to share soothing words, the British authorities did not share extensive information about the true nature of the weapon, in keeping with all other tests in Australia. Such information was shared among the British nuclear elite, in the form of memos and briefing documents and technical reports, but none of that information was provided to Australia. Members of the AWTSC were briefed in somewhat more detail than any other Australian officials, but they kept their thoughts to themselves, while giving the test their approval to proceed. The extent of their knowledge is unclear, since they did not provide any detail in their meeting minutes.

The draft of a message composed specifically for Robert Menzies was included in the CRO telegram. This message, devised by William Cook when determining how best to explain the Mosaic experiments to the Australian government, indicated that the test authorities were seeking to obtain scientific data possible when observing a fission device that includes the same light elements that are used in thermonuclear weapons. Among the elements used in these weapons, apart from isotopes of hydrogen, are forms of the lightest metal element, lithium. These elements were expected to react in the explosion, but the effect on yield would be minimal. The original plan for G2 did not include uranium, as this message indicated: 'The larger yield weapon is a fission weapon containing somewhat large quantities of the light elements, but containing no uranium with which the light elements can react. The behaviour of the light elements will be diagnosed with the help of traces of various materials'.[56] Once G1 had been detonated, and it was apparent that the light elements had made little difference to the explosive yield, plans for G2 changed to introduce the uranium tamper. G1 had shown that doing so would not put the weapon over the 100-kiloton limit that AWRE had promised the AWTSC.[57] The note continued with an explanation of the nature of a hydrogen bomb in which the major explosion results from large amounts of light elements fusing and releasing energy in the megaton range. As agreed by British authorities, the true nature of Mosaic was not to be revealed to the Australian government, just a hazy description: 'The two rounds at Mosaic are of low yield – a few tens of kilotons – and the small amount of light elements incorporated are solely to investigate the nature of the reaction'.[58]

The Australians were clearly either confused or misinformed, because Minister for Supply Howard Beale released press information that indicated that G2 would be the smaller of the two weapons tests. He said in his media release on 27 May 1956, ten days after G1 was detonated, that a 'further smaller device will be detonated in a few weeks [sic] time when conditions are favourable'.[59] Exactly why Beale held this erroneous view and made it public is not entirely clear, although it may be understandable giving the frequently changing plans for Mosaic. Some earlier British planning documents had speculated about switching Mosaic to Maralinga and reducing the proposed yield, one of a range of possibilities that were considered. As John Symonds wrote in his history of the tests, 'whatever the reason for Mr Beale's remark it was incorrect and created some grounds for the belief that "the G2 device went wrong"'.[60] A flurry of media activity arose when a Japanese coal ship, *Miyajima Maru*, recorded radioactivity on board at its location some 400 kilometres east of Mackay in the Pacific Ocean.[61] One media story said that the ship, which had docked in Newcastle, had recorded a high Geiger counter reading of 6000 counts per minute, compared with a normal reading of 15. 'The 6,000 count is three times as high as that reported by a Cloncurry miner in geiger tests of rain water last week', the story reported.[62]

As the G2 cloud was quickly rising and spreading, so too were Australian concerns about the latest British test at Monte Bello. The G2 cloud developed a marked 'bulge' on one side of the stem somewhere between 18 000 and 24 000 feet.[63] This breakaway cloud may have contained about 15 per cent of the debris, and is most likely the main source of contamination as the winds changed direction and sent it towards the mainland. This contamination is probably what was picked up by the Geiger counters wielded by Stewart Stubbs at Marble Bar, and was later detected at Onslow, Dampier, Port Hedland and Broome. The material from the cloud dispersed as far as Queensland. Soon, the media picked up the story and ran with it, in some cases predicting 'death and destruction'.[64] Much later, just before the Royal

Commission got underway in Australia, journalists were starting to peel back the layers of secrecy. An investigative article in *The Times* of London would one day lay bare some Mosaic truths: '[b]ecause there has been so little investigation of the consequences, we may never know how much damage was done, but we have uncovered evidence of immense anxiety behind the official façade of confidence'.[65]

Media reports about the G2 blast added to the sense that something had gone wrong at Monte Bello. One report, in the *Canberra Times*, quoted opposition leader Doc Evatt as saying,

> Mr. Beale's assertions as to the complete safety from radio active effects must be treated as worthless. It is obvious that the whole situation affecting atomic explosions must be reviewed as a matter of extreme urgency. Mr. Beale's statement as to the chemical and physical make-up of the high cloud is merely one more unauthoritative assertion.[66]

Although media reporting on the Royal Commission hearings in London in the 1980s suggested that Robert Menzies sent a message to Eden saying, 'what the bloody hell is going on, the cloud is drifting over the mainland',[67] the evidence suggests that this message was actually sent by Acting Prime Minister Arthur Fadden. Menzies had travelled to London via Malta for the Commonwealth Prime Ministers' Conference at this time, and Fadden had responsibility in his acting role. Royal Navy radio operator Bernard Perkins was aboard *Narvik*, operating the radio communications equipment, at the time of the detonation. He gave evidence at the Royal Commission that he saw a message taken by another radio operator and written on a pad. 'I saw the signal come in. It was written in handwriting.'[68] He testified that it had come from Sydney from the Australian prime minister, addressed to the British prime minister, and that an Australian official on board, possibly Ernest Titterton, was summoned to the radio room when it was received. Whoever that official was had seemed perturbed when he saw the message. When evidence about this prime ministerial message was put to William Penney the day after Perkins' testimony, he professed his 'astonishment'.[69] He said he had not been aware of such a message, although some public airing of it had preceded the Royal Commission. In fact, behind the scenes the Australian government had been jolted by *The Times* of London report on 18 June 1984, picked up in Australian media a few days later, that referenced Fadden's blunt demand to the British government for details about fallout from G2. A letter from WG McGregor, the first assistant secretary, Uranium and Nuclear Division of the Department of Resources and Energy, to the minister, Peter Walsh, on 22 June 1984, just weeks before the Royal Commission began, showed the growing concern at high levels of the Australian government. 'Our knowledge of the Mosaic tests is not complete and we will be investigating these matters further.'[70] McGregor said that while the government would not be able to confirm or deny the allegations about fallout, 'it would be unwise to dismiss them'.[71]

Perkins also gave evidence that a Geiger counter on the deck of the ship 'went berserk' after the G2 blast. Soon after, he said, scientists who had been on Hermite Island for the G2 blast were brought back to the ship 'in shock'.[72] 'They had blankets round them and to get out of the boat they had to be helped out; where normally the scientists came back and they would hop out of the side of the raft and jump they had to be helped up the gangplank.'[73] This account is somewhat supported by a cable sent from Maddock to Roy Pilgrim at Aldermaston, although it is couched in jovial terms. He said 'bang was a terrific sight from H1. Nearly required the Elsan',[74] referencing the brand name of a type of portable chemical toilet. Penney was asked about this incident during his first London testimony, although he was not present at Mosaic. He said he had heard nothing about it.[75]

Acting Prime Minister Fadden issued a lengthy statement on 25 June 1956, providing soothing words and assuring Australians that there had been no risk from G2. He noted that origins of public concern had been uranium prospector Stewart Stubbs's Geiger counter readings at Marble Bar, combined with the grounding of some air services in the area.[76] Fadden dismissed the Marble Bar report, saying whatever the miner had detected 'was completely harmless'.[77] In any case, he said, '[r]adioactive rain is a well known phenomenon irrespective of weapons tests'.[78] Quoting the AWTSC, Fadden stated that the timing of the blast had been chosen because it would ensure that all radioactive particles would be dropped into the sea, and this is what had happened. His urgent message to the British prime minister, and the information he had been provided in response, had succeeded in allaying his concerns.

The Royal Commission put it to Penney that there had been internal criticism of the firing of G2 at all, at huge expense (£3–4 million). Penney bristled, saying 'these were really [William] Cook's experiments, they were organised by him, if there were these complaints I am sure they would have been made to me and they probably would have been defended but I cannot say that'.[79] Penney did not want to take responsibility for the messy and ambiguous nature of Mosaic or its aftermath. By the time of Mosaic, he was starting to become tired of the whole business of building these dreadful weapons at all. He left AWRE in 1959, and from 1962 headed the UK Atomic Energy Authority until 1967, at which time he left nuclear work altogether. He raised his voice against nuclear proliferation in later life. To be hauled in front of a judicial enquiry as the star witness was an indignity that some say may have even hastened his death from cancer in 1991. One of his obituarists, the British diplomat and former ambassador to the United States, Roger Makins, made this claim after Penney died in 1991: 'He was rudely and roughly handled by the Commission, and was deeply wounded and depressed by the experience and by the Commission's conclusions. It is the unanimous view of his family and friends that he never fully recovered from the treatment he received at the Commission's hands'.[80] Penney chose to burn all his papers just before his death.

CHAPTER 10

DIANA, GUINEA PIG SHIP

> What puzzles so many of us is why, on conclusion of the trials, no effort was made to see what effect the nuclear explosions had, both physiologically and psychologically, on the men who witnessed them and perhaps more important, on those who went through the fallout. Yet we had no follow-up medical checks of any kind.[1]
>
> CAPTAIN JOHN GOWER, COMMANDER OF HMS *DIANA*

An air of tragedy haunts HMS *Diana*, one that continues to play out through the generations. *Diana* was deliberately placed into the path of fallout from both Mosaic bombs and its crew received a variety of radiation doses, some high, in the interests of understanding how atomic fallout would affect a ship and those onboard during combat. The safety issues that dogged the whole British test program are brought into sharp focus in the case of *Diana*. The injustice rings down through the years as *Diana* veterans and their families continue to wait for recognition and reckoning. The physical effects of being exposed to radiation have been turbocharged by the psychological effects that came with knowing, after the event in the case of most crew members, that this was a guinea pig ship and its crew were to be unwitting experimental subjects, who did not volunteer to be exposed to radioactivity. The UK Ministry of Defence continues to deny that there was any risk entailed in placing a ship into the path of nuclear fallout. The exact harm caused to *Diana* crew is still a matter of considerable debate, particularly in Britain, and no detailed epidemiological studies have been done specifically on this cohort. Representatives of the nuclear veterans, such as the UK veteran advocacy organisation Labrats, claim high illness and mortality rates for *Diana* crew, based on veteran testimony and a critique of official record keeping. The lived experience of *Diana* veterans and their families versus the denials of the British government powers an ongoing debate that may not be fully resolved because exposure levels on board the ship were not always rigorously collected and the medical records of service personnel have allegedly been actively withheld by the Ministry from many nuclear veterans, not just those from *Diana*.[2]

Diana's role as the fallout observation ship for Mosaic was first mooted at a meeting of MOSEX in December 1955, only six months out from the first trial.[3] Her official task was to 'conduct a number of ad-hoc trials to investigate the radiological hazards caused by fall-out to machinery spaces and machinery'.[4] *Diana* was a member of the Daring class of British warships, designed for endurance at sea. The ship was commissioned in 1954, and not long after that was on her way to the southern Indian Ocean.

Diana's place in history was cemented in two different ways within weeks of each other in 1956. First there was her unique and perilous role in Operation Mosaic, followed immediately by her part in the Suez crisis that brought down the British prime minister responsible for Mosaic, Anthony Eden. In that ill-fated adventure, on 31 October 1956 *Diana*, along with the British cruiser HMS *Newfoundland*, sank an Egyptian frigate, *Domiat*, then picked up 69 survivors from the sea. The ship's commander during both these momentous events, John Gower, had been tapped on the shoulder to take command when its captain at the time broke both legs skiing, just before the ship headed to Australia.[5] Gower published an account of these events in 1995, which is highly readable and also sometimes inaccurate, including calling the Monte Bello Islands 'Christmas Island' several times. His account is crucial, however, to understanding what the crew went through. He emphasised that the aim of placing *Diana* into the path of fallout was not just to understand what might happen to a ship, but also what might happen to those aboard during a nuclear attack. He wrote, 'unknown to me at the time, since 1953 the Chiefs of Staff had wanted to know what effect an atomic explosion would have on ships, their contents, equipment and men ... HMS DIANA and her crew was made available to provide the answers'.[6] In fact, the task assigned to *Diana* was the only ancillary test at Mosaic. Unlike other series, which had extensive ancillary experiments, there were no 'target response' tests to determine the effect of nuclear weapons on equipment during Mosaic. This would be the last opportunity to undertake such a maritime test, since it would not be possible at the new permanent site at Maralinga. In their book on the British tests, authors Arnold and Smith identified *Diana*'s deployment at Mosaic as the 'naval counterpart to the RAF's *Operation Hot Box* at Totem in 1953',[7] during which aircrew were deliberately exposed to radiation from the first Totem test. In Hot Box, however, only three aircrew were in the RAF Canberra that entered the cloud six minutes after detonation, and all were thoroughly familiar with the risks. The same could not be said for the crew of *Diana*.

Diana was one of four ships identified by the Admiralty as being suitable for the Mosaic assignment. The others were Her Majesty's Ships *Daring*, *Delight* and *Decoy*.[8] The selection of *Decoy* and *Diana* would cause the least disruption to naval operations, although there would still be some. A meeting of the British vice chief of the Naval Staff in December 1955 noted that the provision of *Diana* 'is at the expense of the Mediterranean Fleet and in order partly to compensate the Mediterranean Fleet for this loss, DARING should remain in the Mediterranean until about mid-June'.[9] Three objectives were identified: to collect scientific data on fallout; to test gear and arrangements made; and to 'indoctrinate' (in the sense of instruct) officers and men in the hazards of fallout.[10] *Diana* and her crew were to be purposefully placed directly into the danger zone, so that the effects of an atomic attack could be observed first-hand.

A few weeks earlier in 1955, the Admiralty had acknowledged the unique opportunities afforded by Operation Mosaic, should suitable vessels be available, and set out two objectives:

> <u>Object One</u>
>
> To exercise ships in the procedures laid down for protective action under atomic attack. This will include pre-wetting the ships before arrival of fall-out, and measurements to determine its effectiveness.

> Object Two
>
> To collect fall-out samples for measurement and subsequent analysis and to determine the rate of vertical diffusion of fall-out in the sea.[11]

The original plan was to position two ships, at distances of 322 and 644 kilometres respectively from ground zero, for both Mosaic tests. 'At neither distances on either occasion is it considered that harmful radiation doses can be received even supposing the methods of protection adopted are ineffective.'[12] In the end, only *Diana* fulfilled this role, however, and she was considerably closer to the tests.

While it was not possible to deploy two ships for the task, there was a fallback plan. An unmanned scientific whaler filled with all the same measuring apparatus as aboard *Diana* was intended to obtain closer-in data. This small, open boat was to be dropped off by *Diana* before she took up her own position ahead of receiving the fallout. Distances had to be calculated using estimates of wind speed and the expected yield of the weapons, with the yield range used for these estimates being 5 to 120 kilotons.[13] Planning documents noted the difficulties obtaining the yield part of the calculation, since yields were expected to change right up to the time of firing.[14] The whaler was intended to mimic the fate of a ship caught by surprise by an atomic attack.[15] The distances calculated for it reflected the expectation that casualties would result from being in the path of fallout. It was planned to be positioned closer to the ground zeroes than the Admiralty dared to place a crewed vessel. After dropping off the whaler, the captain of *Diana* was to move as quickly as possible away and take up its own position at a distance that would only be known when other variables such as wind and yield had been factored in. For G1, *Diana* was 64 kilometres north of the tower on Trimouille Island, the opposite direction from the other ships of the fleet that were positioned in safe places to the south. For G2 she was 156 kilometres north of Alpha Island. In the event, the whaler was only able to record data for G1, since it was sunk the day before G2. For G1 only, it was positioned 9 kilometres from ground zero. *Diana*'s commander was instructed to take off at speed if the readings got too high.

> If, in spite of all precautions against contamination from fall-out, DIANA should find herself in unpredicted and excessive fall-out, complete shelter stations are to be ordered and the ship is to steam at best speed in that direction at 45 degrees to the bearing of ground zero which is most nearly opposed to the surface wind.[16]

The major safety precaution to be tested was 'pre-wetting', in which water was sprayed continuously over many parts of the ship, actions designed to quickly rid the ship's upper decks of fallout. The Americans used a similar system, which they called 'washdown', for their Pacific tests, such as Operation Castle in 1954.[17] A report prepared by the Admiralty after Mosaic asserted that pre-wetting worked exceptionally well, reducing contamination by 80 per cent.[18] Evidence for this claim came from two instances, one during G1 and the other during G2, when the pre-wetting system had not worked for various reasons in certain parts of the ship, and those areas were found to be significantly more contaminated than areas subject to continuous wetting.[19]

A restricted report published for a conference on the effects of atomic weapons in London in 1957 noted that during both Mosaic G1 and G2, 'DIANA received just about the intended amount of fallout'.[20] Extensive planning was undertaken to ensure that *Diana* would be in the right place at the right time, although many of those calculations proved to be inaccurate. An AWRE report by Albert Matthewman on meteorological preparations set out the role of weather prediction in positioning the ship. The report noted that one of his group's responsibilities was to ensure that *Diana* was located 'at a place where the predicted dose would accord with her scientific function and with the precautions available to her'.[21] More than 300 officers and men were on board the ship for the operation.[22] They endured cramped quarters, high temperatures and humidity, a prohibition on smoking, limited supply of hot and cold drinks, and monotonous food rations.[23] Nevertheless, morale was reported to be high.

Operation Mosaic was seen as a golden opportunity to more fully understand what would happen to ships during an atomic war. The scientific tasks *Diana* was required to perform were under the direction of the Admiralty, not AWRE.[24] The orders issued for *Diana* for Mosaic G1 spelled it out: 'Operation MOSAIC is the only likely chance in the near future of testing the protection of modern ships when passing through the fall out of a nuclear explosion'.[25] *Diana* was to be present for both Mosaic bombs. She departed from Britain on 29 March 1956 and anchored in the Monte Bello parting pool on 2 May. She had made a fast passage from Britain to be at the islands in time for the G1 rehearsal, originally scheduled for 4 May but postponed to 5 May.[26] She would remain anchored in the parting pool when not out at sea; her draught was such that she could not enter the lagoon.[27] This issue caused some problems, since *Diana* did not have its own power boats for moving between vessels, and had to rely on those of *Narvik*.[28]

Diana's scientific whaler boat was slipped over the side at 6 am on the day of G1 roughly 9 kilometres from Trimouille Island. *Diana* was then ordered to steam to her trial station to the north.[29] Among the gear on the whaler was sticky paper to pick up radioactive particles, as well as a small radar antenna to enable the vessel to be located and collected. Plans for G2 did not go smoothly: the whaler had 'inadvertently' overturned and sunk the day before during a false alarm for the firing.[30] It had to be replaced with a buoy that was equipped with sticky tape and collector trays to gather radioactivity.[31] Once the buoy was deployed 11 kilometres from ground zero, *Diana*'s commander was ordered to position his ship such that she would receive fallout 'from the bottom of the puff of the atomic cloud'.[32] A later report found that this would have gone 'extremely badly' if the commander had obeyed this order – only in the sense that the ship would not have received enough fallout for the purpose of the exercise.[33] Instead of waiting passively for the arrival of the fallout, the ship steamed right into the thick of it, intercepting the falling radioactive particles about eight hours after detonation. 'Had the ship waited, the activity would probably have been down by a factor of 20 or so on what was actually obtained.'[34]

The observed dose rates on the deck of *Diana* after G1 were higher than predicted, possibly because the ship was being washed down with contaminated sea water.[35] Later analysis of data from the scientific whaler after G1 showed that fallout started arriving at its location faster than predicted, only 25 minutes after detonation. For G2, without the scientific whaler but using the buoy 11 kilometres from ground zero, fallout from the test started arriving about two and a half hours after detonation.[36] The fallout report made the point that safety precautions on *Diana* may indeed have been too strict, since no one on board received a measurable gamma dose, even though the upper deck was greatly and purposefully contaminated. The report speculated that 'one important cause of this happy result was undoubtedly the fact that the ship was fairly thoroughly decontaminated within a few hours of the arrival of the fallout'.[37] Captain Gower noted in his own report that when canvas covers were removed from the bridge and elsewhere on the ship soon after the fallout had stopped arriving, the crews found all covers had radioactivity readings in excess of 10 milli-roentgens,[38] a relatively low dose but still significant. By early the next morning contamination readings on the upper deck had fallen substantially. Gower reported, though, that some 'were declared by the Scientific Staff to be giving unacceptably high counts'.[39] These positions were roped off for a up to 20 hours until after repeated decontamination before they were declared safe to use.[40]

Once G1 was detonated, the crew were to use 'doll codes' to indicate the time of arrival of fallout, and estimates of dose rates on various parts of the ship – pre-wetted and unpre-wetted exterior surfaces and between decks.[41] The British used multiple codes during Mosaic, including the 'doll codes' that were used to indicate data connected with fallout and health physics in communication between HMS *Narvik* and HMS *Diana*. These codes were: Alice (1), Mary (2), Carol (3), Anne (4), Jane (5), Kate (6), Jill (7), Dora (8), Emma (9) and Fay (0).[42] Using these codes, it was possible to indicate, for example, time of arrival of fallout at 1930 hours as Alice Emma Carol Fay. The codes were intended to provide a simple means by which Hugh Martell aboard *Narvik* could know exactly when *Diana* began experiencing fallout.

The British were still relying on the same safety standards set for Hurricane, including their unique measuring system for doses of radiation. The trial orders set out safety standards for *Diana*'s crew, differentiating between lower integrated dose (LID), higher integrated dose (HID) and special higher integrated dose (SHID). LID could be applied to any crew member during each trial without special permission. HID, however, required the approval of *Diana*'s health control officer and medical officer. SHID required the permission of Commander Hugh Martell himself. In the case of both HID and SHID, '[m]en to whom this dose is applied are not to be exposed to further radiation above L.I.D. for the remainder of the operation'.[43] As they had for Hurricane, the orders noted that if 'it should be necessary to apply higher integrated dose to any personnel, older rather than younger men should, if practicable, be chosen'.[44]

An extensive report was prepared by the Director Tactical, Ship Requirements and Staff Duties (DTSD) at the Admiralty to guide Martell on how to position *Diana* during the operation. The Admiralty briefing contained a variety of safety information as well. Handwritten by the civil servant who provided the report on behalf of DTSD, Captain Guy Hawkins, are the words: 'Even if the whole ship's company were exposed to the "special higher integrated dose" of 25r it is unlikely that any casualties would occur'.[45] The implication was that the Admiralty was prepared to take that risk during this experiment. The safety standards that applied to the rest of the force at Monte Bello for Mosaic were eased when it came to *Diana*.[46]

Although *Diana*'s orders came from the Admiralty, the AWRE sent its health controller, James Hole, to inspect *Diana*'s safety precautions. He suggested a number of tweaks to the safety arrangements, especially since the ship was not going to lock down completely while in the fallout zone and people would be moving about. Because of this, various precautions were ordered, such as improved clothing and respirators for people working in the boiler room, as well as the provision of dosimeters.[47] Appended to this report was a note from AWRE scientific supervisor Ieuan Maddock to Commodore Martell, who cautioned

> Mr. Hole is only expressing his own opinion as an expert in this field, and the offer to lend equipment and clothing is being made to assist DIANA carry out this exercise. We can not, however, assume any direct responsibility for the Radiological Safety of DIANA since this lies outside our terms of reference'.[48]

Safety responsibility lay with the Admiralty, placing *Diana* into something of a grey area.

Keith Syder was a young able seaman aboard *Diana*, aged only 19. In his statement to the Royal Commission, he recalls being told that his ship would be involved in the bomb tests only when they were on their way to Fremantle, a familiar story about the way the lower ranks were routinely kept in the dark about their atomic missions. When in the Western Australian port, the ship took on board 'paravanes', a kind of underwater glider dragged on a hawser from the ship, loaded with equipment for recording radioactivity.[49] Scientists also joined *Diana* at Fremantle. Syder, so young and completely inexperienced in anything to do with atomic weapons, said that he 'had a very matter of fact attitude to the whole business and was not particularly impressed by the fact that I was about to witness the tests'.[50] He was so relaxed about it all, he could only clearly remember one of the tests, although *Diana* was present for both Mosaic blasts. 'We were wearing battle dress, that is blue denim shirt and blue jeans. We were given a set of sea boots, a white heavy cotton hat, a light overall made of cotton or plastic and I think we may have had a mask over our faces.'[51] Almost immediately after detonation, *Diana* began sailing towards the cloud, 'which was very obvious and seemed to be descending towards the sea ... We had to wait until the fall out cloud was visible before we began sailing straight into it'.[52] He remembers some personnel, most likely scientists, who were wearing white 'spaceman' suits. In his evidence to the Royal Commission, he made the point that the 'scientists were definitely dressed differently and more adequately'.[53]

As they sailed through the cloud, Syder noted that it 'seemed to be made up of fine brown dust particles. The cloud was browner than fog and not so thick. The sunlight was dispersed or refracted through the cloud'.[54] He expanded on this description when giving evidence, saying that it looked like 'a dust haze across the sun's rays ... It was almost like a very light dust storm'.[55] In later life, Syder developed a skin condition and growths in his mouth that his dentist could not identify.[56]

Diana was refused entry to the port of Fremantle on her way to Singapore, owing to Australian government concerns about contamination. Refuelling *Diana* was challenging, since she was going to be deliberately contaminated. Commander Martell noted in his report, 'it was considered undesirable to send her to an Australian port in case political capital might be made from the fact that the ship had been contaminated, however slightly'.[57] After G1, HMS *Alert* had to transfer 100 tons of fuel to *Diana* to enable her to get to Singapore.[58]

Kenneth Stephens was also a young British seaman aboard *Diana*, who moved to Australia to live in the early 1960s. He remembers being given some fairly vague and generalised lectures on atomic testing before leaving Britain, as part of a 'damage control school' he attended in preparation for managing the ship's damage control system.[59] Only when they were at sea did they learn of the mission they had been sent to undertake, although then just in outline. During the blasts he was kitted out in normal clothing, but with the addition of 'a plastic style oil-skin coat, leggings and boots. We had a hat but I cannot recall what style it was. We looked very much like north sea fishermen'.[60] He and his colleagues had also been issued with radiation detectors 'and I was sufficiently aware to understand that their presence indicated that there was some risk'.[61] Stephens' role was to test the effectiveness of the pre-wetting system. 'I think we probably looked like a celebrating New York tug which had turned its fire hoses to the air,' he commented.[62] When the crew's radiation detectors were examined on the ship, Stephens noted that people receiving the detectors exclaimed how high the readings were. 'This is why I think that they were taken from us because they were being read as we handed them over. I did not know how to read them and the relative values of the reading would have been lost on me.'[63] He says he also saw some of the personal dosimeters being thrown overboard. He was involved in sampling water, most likely using a paravane (although he did not name it as such in his statement). 'I accept that this indicates that we were steaming through radioactive water,' he said.[64]

Howard Coles was aboard *Diana* for both Mosaic tests, undertaking the scientific evaluation of the pre-wetting sprinkler system. He was required to monitor the levels of radiation within hours of detonation. When much of the crew was confined to the interior of the ship, Coles and his colleague were out on the deck collecting data about the fallout that had been deposited on the ship.[65] His protective gear was limited to anti-gas oilskins, an anti-flash balaclava and a gas mask that had to be removed when he needed to convey Geiger counter data to central control. 'We were obliged to remain on deck during the period of monitoring in which certain sections of the ship were deemed to be "hot".'[66] He reported that his personal decontamination routine was 'perfunctory', involving just a shower and a once-over with a Geiger counter by a non-medical person. 'No further "medicals" ensued,' he said.

In 1956, Ronald Scott was a young engineering mechanic on board *Diana*. He testified later at the Royal Commission that he had watched the first Mosaic explosion from the upper deck and within an hour he and his shipmates had gone to their 'ABC' stations, standing for 'atomic, meteorological and chemical warfare stations' below deck.[67] There they stayed for between 16 and 18 hours as the ship steamed through the cloud of radioactivity. The same procedure was followed for the second test. He said, 'we were ... just steaming probably in circles at about 10 or 12 knots'.[68] The Admiralty presumably was concerned about the dose that would be received by crew below deck, since Scott testified that two crew members continually took Geiger counter measurements during this time.[69]

Australian naval officer Arthur Andrews had a unique perspective on *Diana*. He was the only Australian on the ship and his presence at Mosaic was requested by the Australian Commonwealth Naval Board.[70] Andrews had previously headed the Joint Services Training Unit that operated at Monte Bello after Operation Hurricane. He was to come back to Australia in 1956 as an observer on board *Diana*, though not to the islands themselves as the ship stayed out in the fallout zone. He worked with the senior scientific officer, a Mr Thomas, on safety measures such as pre-wetting and monitoring seawater coming through the ship's water inlets. He was especially interested in the pre-wetting protocols established for *Diana* – 'we considered this to the best method by which a ship could be protected in the event of an atomic attack and that is what we wanted to know all about'.[71] He professed himself 'delighted' that radiation was detected on the ship, 'because no-one knew, not even the Americans after tests [Able and Baker at Bikini Atoll] ... exactly how much radiation one could remove from the ship'.[72] His measurement of the radiation on the ship was initially quite high, but it fell away sharply – 'it had diminished by at least three quarters, which was something we had not expected'.[73] While there were some hot spots where the pre-wetting had been less effective, he said that the safety measures were better than he had been expecting. As a good navy man, he lauded the superior naval safety standards, which stood in contrast, he said, to army and air force operations he witnessed later at Maralinga. 'I do not think you will find many naval people on your list of people who suffered from radiation because we were perhaps rather more rigid in what we did and said,' he stated.[74] Symonds claims in his history of the tests that Andrews 'was later commended for his work on DIANA, particularly after the first test when he suggested alterations to DIANA to improve decontamination procedures – these were carried out at Singapore during a visit between trials'.[75]

Diana's diversion to the Suez crisis detained her for three months, at Aden in the Middle East. Upon the ship's return to Britain, Captain Gower was astonished to find that no medical investigations were to be done on the crew, even though they had sailed through the fallout of two atomic tests.

> Why not get value for money by testing everything – men as well as weapons? They did it in the United States. The answer given by the Prime Minister [Margaret Thatcher] in 1983 was that no one was exposed to any significant health hazard; therefore no tests were necessary. But without testing, how is it possible to be sure there were no hazards?[76]

Campaigning journalist Susie Boniface has been one of the leading British figures attempting to get the UK government to accept responsibility for the harms caused by British nuclear tests. As part of an ongoing investigative series, she documented the case of Bob Malcolmson, who was aboard *Diana* as a teenager, working as a wireless operator. Four years after *Diana*'s voyage to Monte Bello, Malcolmson was diagnosed with a bone marrow disorder that was later admitted by the Ministry of Defence to have been caused by exposure.[77] In 1990, he was compensated with less than £2500. The British government, generally unwilling to admit to any problems caused by the tests, concedes that the crew of *Diana* were among those exposed to 'measurable levels of ionising radiation'.[78]

CHAPTER 11
AIR OPERATIONS

> We were not given any blood tests prior to that flight nor was I given blood tests or any other health checks in respect to radioactivity after that flight. To the best of my knowledge, my aircraft did not receive any decontamination treatment nor was it washed down at Broome. Approximately 12 months ago [in 1983] a blood count disclosed that my blood was not quite in order and further checks were made. As a result of those checks I have been informed that I have lesions in the bone marrow.[1]
>
> ERIC McHARDIE, RAAF NO. 82 WING, CAPTAIN OF LINCOLN A73-41, ONE OF THE FIRST PLANES TO FLY INTO THE HURRICANE CLOUD ON 3 OCTOBER 1952

While both Hurricane and Mosaic had a distinct maritime flavour, the two operations also had extensive aerial aspects. One of the most dangerous roles for military personnel working at the tests was cloud flying and tracking. Brave airmen from Britain, Australia and elsewhere were tasked with taking airborne samples and tracking the course of the long-lasting, wandering, often elusive clouds. The clouds carried radioactive particles far from the site of the blast, in directions governed by their height and the prevailing winds. The AWRE produced some safety information, but much of the testimony of cloud flyers later suggested that they had little understanding of the full range of risks to which they were subjected. Many did not recall any safety measures or training, especially in relation to Operation Hurricane. Indeed, Hurricane stands out for its almost complete lack of safety standards for air and ground crew. At least some of the cloud tracking was primarily concerned with providing the authorities with plausible statements about the safety of the test, rather than obtaining evidence of the passage of a radioactivity-laden cloud over the Australian mainland. Orders that could have determined the actual path of radioactive clouds were not always issued. Cloud tracking also appears to have been carried out for intelligence reasons, to provide more information on how the scientific characteristics of atomic bombs could be analysed from a distance by friend or foe.

Australia's contribution to the airborne aspects of Hurricane was relatively modest but still significant. In many ways, RAAF involvement in Hurricane was as much for political as for operational reasons, providing the appearance of a partnership between Britain and Australia that did not really exist. Having Australian military planes in the air did serve a potent political role, one that helped reassure Australian sensibilities. The cost was later estimated to be £67 000 of expenditure on RAAF services, excluding pay and maintenance costs,[2] the equivalent today of around $3 million. The backbone of the RAAF contribution to Hurricane was 12 Lincoln aircraft and their crew, based on both the east and west coasts, out of Townsville and Broome, supplemented by several Dakotas used for radiation surveys and courier duties. As AWRE scientist Noah Pearce commented in his statement to the Royal Commission in the mid-1980s, 'provision of aircraft by the Australians was most generous'.[3] A smaller amount was committed to Mosaic, but still substantial given that that operation had not been budgeted and was thrust upon the Australian government with little warning. For Mosaic, a Canberra crew (who operated an RAF aircraft) and a Neptune detachment from No. 11 Squadron Richmond operated out of Onslow. The Canberra crew flew into both Mosaic clouds. The Neptunes were primarily tasked with security patrols.[4]

The RAAF had a diverse range of roles at Monte Bello, from flying sampling and tracking sorties and security operations to undertaking infrastructure and engineering activities. For Hurricane, the RAAF sent detachments from No. 82 Wing based in Amberley in Queensland, No. 86 Wing from Richmond in New South Wales, and No. 5 Airfield Construction Squadron from Williamtown, New South Wales. The detachment from No. 82 Wing was positioned in Broome, while that of No. 86 Wing was based at Onslow. The North Queensland end was co-ordinated from RAAF Base Townsville (Garbutt).

No. 86 Wing arrived in two groups, the first of two aircraft in August 1952, then the second in September, to carry out security patrol duties out of Onslow.[5] Around the same time, No. 82 Wing arrived in Broome, to begin air-sampling operations. A tented base camp was established on the grounds of a local meatworks, and the hectic, frontier Broome nightlife, along with dances organised by the locals, proved to be popular with personnel.[6] Air Commodore William Hely, deputy chief of the Air Staff for the RAAF, wrote to RAAF Headquarters Eastern Area after the Hurricane test, saying that he had inspected the Broome detachment and 'I am able to report that the morale of the Detachment was high, and its relations with the citizens of BROOME were excellent and that the people of the town were very sorry to see the Detachment depart'.[7] There was rather more fun to be had in Broome than in Onslow.

No. 5 Airfield Construction Squadron, also known as 'the flying shovels', had a short but proud heritage of building airfields and other infrastructure since the Second World War (and would build the atomic test site at Emu Field in 1953). They worked closely with the British Royal Engineers, constructing many facilities essential to the operation such as roads, instrument foundations and light piers, and also helping to install shark-proof nets in swimming areas.[8] When the scientists finally arrived, they found a remarkable amount of infrastructure had been created, where none had existed only a short time before. The 'flying shovels' had a formidable reputation as hard, efficient workers who moved mountains in often harsh conditions.[9]

As Hurricane planning got underway, the British authorities sought New Zealand government assistance because the radioactive cloud was expected to go beyond the east coast of Australia and out into the Pacific. While the RAAF would monitor cloud movements over the Australian continent, the British government wanted help tracking the cloud even further east. A telegram to the UK high commissioner in New Zealand in May 1952 indicated that the meteorological data suggested that the invisible cloud was expected to pass near Fiji, at which point it would have a horizontal radius of about 482 kilometres.[10] A cloud of that size would be significantly larger than most tropical cyclones. New Zealand agreed to undertake this monitoring, and aircraft flew from Whenuapai, near Auckland, collecting considerable radioactivity.[11]

The Western Australian coastline between Onslow and Broome, a sweep of more than 1100 kilometres, was known to be vulnerable to radioactive contamination from Operation Hurricane.[12] The area was surveyed by air with gamma sensitive detectors on RAF Dakota aircraft ahead of the first British blast. This was to be the first time that the remote Western Australian coastline was scanned from the skies as part of the atomic tests. There were many more sorties to come.

During Operation Hurricane, RAAF aircrew flying unpressurised Lincolns out of Broome – the first aircraft to enter the Hurricane cloud – were essentially unprotected. The Royal Commission found that they did not carry film badges or dosimeters and the aircraft had no gamma-ray detection equipment. 'The only radioactivity collection systems were the filter units attached beneath the wings of the Lincoln aircraft.'[13] The outer surfaces of the aircraft became contaminated with radioactivity following sorties into the clouds. Nevertheless, estimates accepted by the Royal Commission of the dose received by aircrew were low.[14]

> These estimates make it likely that the surface contamination on the Lincolns gave very low dose rates and that the maximum dose to the aircrew was well below the normal working dose specified in the Hurricane Radiological Safety Orders. The contamination levels were such that ground crew working on the aircraft after their return would only have received radiation exposures that were well within the approved lowest radiation dosage level for Hurricane participants, called 'the Normal Working Rate'.[15]

Despite this statement, elsewhere the Royal Commission also notes that RAAF aircrews had no dosimeters and were outside the safety orders, rendering their doses 'impossible to determine accurately';[16] moreover, procedures to check and decontaminate aircraft were lacking, despite evidence of contamination and subsequent segregation and steam-cleaning of Lincolns and decontamination effluent issues at RAAF Base Amberley.[17] Ground crew frequently had more exposure than aircrew because of the nature of their work, clambering over aircraft on the ground to repair or refuel them.

British scientist Hubert Gale was charged with analysing the filters from aircraft carrying out sorties in and around the Hurricane atomic cloud. He later speculated on the various possible purposes for these operations. Gale was based at AERE at Harwell and reported to Dr William Marley, head of health physics. There, Gale was part of the Atmospheric Dissemination Group and involved in detecting radioactivity from the first Soviet atomic test in 1949. His role had been to help determine if the radioactive material collected over the Atlantic was due to a reactor accident or a bomb. Analysis quickly revealed that it was a bomb.[18] This Harwell work, driven by the need to monitor Soviet atomic testing, helped Gale's division to hone its skills. He found himself drawn further into analysing radioactivity gathered from filters attached to aircraft, and this prepared him for his role in Hurricane, and later Operation Totem. In his statement to the Royal Commission, he said that at some stage in Hurricane planning 'it was decided that cloud sampling operations at long distances would be carried out at points removed from the actual site of the explosion'.[19] He believed that this was mainly for intelligence purposes, to provide a control against which the results of other countries' tests could be measured, rather than to determine the efficiency of the British bomb: 'the tests necessary for that purpose were carried out by rockets. This however was not known to me and I do not believe I was fully aware of the purpose of our experiments in 1952'.[20] Gale's ponderings were bolstered by evidence from William Penney. When it was put to him by Peter McClellan that one of the main reasons for cloud sampling in the British nuclear test program 'was to inform the United Kingdom about methods of analysing the Russians' or the Americans' explosions', he agreed.[21] He said that obtaining samples was necessary to test methods for determining the nature of bomb blasts; by sampling their own weapons they could gauge the accuracy of their detection methods. Gale speculated that this was why the cloud-sampling operations became separated from the main Hurricane planning, and safety fell down the list of priorities. The Royal Commission found that the air-sampling effort branched off well before Hurricane and was run separately to the main operation, with only a 'tenuous connection' to the Monte Bello command structure.[22] 'By May 1952, administrative arrangements for the air sampling operation were being made by Air Vice Marshal Edward

Davis in the Ministry of Supply. He was concerned that the operation was lacking co-ordination and began to question the Naval Commander about the movement of people and equipment.'[23]

Gale set up a lab at Broome to receive and analyse the filter samples from the cloud-flying aircraft. 'At no time during the preparation for Hurricane was I aware of any discussion on the possibility of aircraft or personnel being contaminated,' he said in his statement.[24] He was asked to assist aircrews to locate the cloud. In Broome, five RAAF Lincolns were fitted with filters to sample at 800 kilometres from the test site, while in Townsville another five RAAF Lincolns were likewise fitted with external filters, to sample at 3200 kilometres away. In Townsville, 10 (GR) Squadron was under the command of John Hampshire, who reported to the air officer commanding the North Eastern Area, Air Commodore Ian McLachlan.[25] Hampshire would later take part in Operation Buffalo at Maralinga. Because 10 Squadron did not have enough aircraft for the needs of Operation Hurricane, three Lincolns and their crew from 82 Wing were attached.[26] Hampshire recollected later in his statement to the Royal Commission that he had little to do with the British scientists, and they did not brief him on radiation hazards.[27] He also noted that no decontamination procedures were carried out on the Townsville-based aircraft.[28]

The Townsville crew were put on 24-hour notice, to carry out sweeps at any time of the day or night at heights of up to 25 000 feet from 24 September, in anticipation of the firing.[29] They flew sorties 44 hours after Townsville base received word of the detonation on a civil broadcast news bulletin.[30] The first two flights came back with no radioactivity in their filters, but the following three flights more than 60 hours after detonation did collect significant radioactivity, particularly in the sector between Townsville and Rockhampton.[31] The samples were collected and some were sent via Darwin to Britain for analysis, while others were sent to Melbourne, under the direction of AERE scientist Raymond Crooks.[32] Air Vice Marshal Davis had earlier sent explicit orders to Gale and Crooks that if radioactivity was found on aircraft flying out of Townsville or Broome, that 'aircraft be monitored externally paying special attention to recesses where lodgement might occur. Information so obtained will assist Air Ministry in formulating aircraft cleansing procedure after flight through similar clouds'.[33]

Squadron Leader Noel McPhail was senior air staff officer North Eastern Area, based in Townsville. His concise but comprehensive report on the Townsville operations was attached to the statement made to the Royal Commission by John Hampshire. McPhail was in touch with the RAAF commander in Broome and the *Campania*, and was tasked with manoeuvring the Queensland-based RAAF aircraft in response to requests from the command ship.[34] The first signals he received from Broome on 3 October indicated that cloud-sampling activities on the west coast had been successful.[35] The next day he received a signal from *Campania* that the cloud height was 10 000 feet but its position was uncertain. A further signal on 5 October 'intimated that the cloud was proceeding seawards, and a search from Townsville may be impracticable'.[36] Given the lack of specific information from *Campania*, and the vagueness around the exact location of the cloud, McPhail improvised. He despatched two aircraft to try and find the cloud, one of which ventured as far south as Coffs Harbour about 1400 kilometres away. Neither sortie detected any radioactivity. In the absence of any further detailed information from *Campania*, McPhail ordered more aircraft into the air. As the aircraft were taxiing for take-off, he heard from *Campania* that the radioactive cloud was 'stagnating in the Monte Bello area' and that the attempts to detect it by aircraft flying out of Townsville should be abandoned.[37] Undeterred, McPhail allowed the aircraft to continue on their long flights to the south. As the pilots started their return to Townsville, adverse weather conditions meant that one was diverted to Amberley and the other to Mackay. Once they were able to, they returned to Townsville. In his report, McPhail revealed that the 'filters on both aircraft when checked proved to be radio active'.[38] They had encountered the Hurricane cloud. McPhail advised both *Campania* and RAAF Headquarters accordingly.

The sorties continued: one aircraft flew from Townsville to Coolangatta and back, and its filters were also found to be radioactive. Another flew north towards Cairns, but was recalled as its usefulness to establishing whether the cloud had passed south of Townsville was doubted. In his report, McPhail concluded that the operation at his end was a success 'in spite of the plotting organisation on H.M.S. CAMPANIA not being able to forecast the movement of the cloud over the continent after the event'.[39]

Clearly, *Campania* had lost track of the Hurricane atomic cloud. The long delays in receiving signals from Western Australia meant that McPhail and his team were left in the dark during crucial parts of the operation. He noted that the small staff of the operations room in Townsville – two officers and a controller – 'worked long hours conscientiously and willingly. Their powers of endurance being sorely tried'.[40] Squadron Leader McPhail's son, also Noel, recalled that in later life his father had told others that these sorties indicated that the radioactive cloud had been detected over the Queensland capital, Brisbane.[41]

The extraordinary recklessness in sending aircrew into an atomic cloud without protection seems to have had its beginnings in a miscommunication. The Royal Commission took evidence that the head of the Health Physics Division at AERE Harwell, Dr William Marley, had provided advice about the risks that had included the line, the 'radiation dose received by plane crew while actually flying through the cloud from an atomic bomb explosion is expected to be well below lethal level at times later than a quarter of an hour after burst'.[42] The advice continued, saying that contamination of the aircraft was as dangerous as direct exposure, although the breathing of contaminated air while passing through the cloud at speed was less risky so long as it was more than 15 minutes after the burst. The advice concluded: '[t]he simple rule that aircraft must avoid flying through the visible cloud following an atomic bomb explosion is sufficient to guard against injury and at the later times when the visible cloud has dispersed there will be no danger to aircrews'.[43] However, the information actually provided to the RAAF was rather less detailed and nuanced. It simply said: 'The radioactive hazard to aircrews in flying through this cloud is negligible and there is no fear of the aircraft becoming contaminated'.[44] The Royal Commission report stated that the 'origin of this statement is not clear, but it was presented at a meeting, chaired by the Naval Commander, by a senior Ministry of Supply officer with RAF connections who should have been aware of the Marley conclusions'.[45] The RAAF officials did not receive qualifying information, such as how long after the detonation it was safe to fly into the cloud. As a direct result, they sent aircrew into the cloud without due attention to safety. None of this was helped by the muddled lines of command for the air operations.

After Hurricane was detonated, a total of nine flights left from Broome to collect samples in the hours just before and after the explosion. All returned with radioactive material in their filters.[46] The air-sampling requirement was varied on the day of detonation and other modifications to the plan had to be made on the run. Aircrews were recalled for further sampling three days after detonation, duties that were completed the following day.[47] The filters from Townsville had significantly more radioactivity than those from New Zealand, but those out of Broome had radioactivity a thousand times what Hubert Gale was used to seeing at Harwell.[48] He qualified this by saying that the Harwell samples he normally worked with were exceedingly small. Gale said that he carried out his work in Australia without any protective clothing apart from a lab coat, and no dosimeter. In addition to the Australian-based cloud flyers, two Royal New Zealand Air Force Bristol freighters and one Douglas DC3 based at Whenuapai, near Auckland, were fitted to sample at 5600 kilometres from Monte Bello.[49] Of the four New Zealand-based sampling flights on 8 and 9 October, three came back with radioactivity.[50] One of Gale's colleagues, Raymond Crooks, was stationed at Townsville and collected filters from the Townsville and New Zealand aircraft. Gale and Crooks went on to author the AWRE report T88/54 'The Collection of Radioactive Cloud Samples by Aircraft Sweeps' in September 1954.

Gale was interviewed by the Royal Commission in London, filling in a lot of the detail on the planning and execution of cloud-sampling sorties for Operation Hurricane. He testified that the radioactivity he was measuring from the filters on the planes would have provided an accurate guide for what the crew were receiving inside: '[t]he filter is acting as somebody's lungs if they were actually breathing it'.[51] Gale reflected poignantly on the marked lack of safety precautions for air or ground crew, or for anyone handling radioactive materials:

> to be absolutely sure of being able to prove it twenty or thirty years later, which is what we have in this current situation, it would have been nice to have the precautions there and the measuring equipment so that you would be able to say 'Yes, there was no hazard'.[52]

The British were doing their nuclear testing on the cheap and in a hurry, though, and with scant thought for the future. They were not to know that the Australian government, so obliging and unquestioning at the time, would one day want to place their decisions under the judicial microscope. Participants such as Gale and many others had to cast their minds back to a time when knowledge was somewhat more limited, pressures were intense and safety was well down the list of considerations. The discomfort this caused in the witness stand is often palpable in the Royal Commission transcript.

Post-explosion surveys one day after Hurricane showed a part of the coast, near Broome, had experienced a localised increase in atmospheric radioactivity at 10 000 feet.[53] Data from these survey flights indicated that levels were nearly ten times the pre-explosion levels.[54] This report was prepared by Douglas Peirson, a Harwell-based AERE scientist who led the radiological survey team for Hurricane. Peirson later gave evidence at the Royal Commission, saying that he had only six weeks' notice before his Hurricane role in Australia.[55] He had been briefed before Hurricane that no fallout was expected to be deposited on the mainland.[56] His work would involve verification, not prevention.[57] He and his junior AERE colleague, Frank Hale, stayed at the tented camp at Onslow from mid-September, and he had good rapport with the nine RAAF aircrew and 20 ground crew, plus a few army members, with whom they shared the accommodation.[58] Peirson and Hale, in separate Dakota aircraft, had to make four-and-a-half-hour low-altitude survey flights up and down the coast. At 6 am the day after the detonation, two RAAF Dakota pilots were ordered to fly at 500 feet above the ground to carry out this survey. Flying this low in hot weather over desert and scrubby vegetation was extremely turbulent, and the pilots were not able to hold this level for the hours they were required to undertake the survey. The flights were uncomfortably bumpy the whole way.[59] The pilots, along with Peirson and Hale, were exhausted; the two AERE scientists had to lie down on the floor of the aircraft through most of the flight. Before the detonation, the pilots had flown practice missions to measure natural background radiation, to establish a baseline, plot a track for navigational purposes and test communications with *Campania*. Later in the day after the survey flight, the scientists and the pilots were back up in the air to track the cloud, flying the same route as for the survey. During this flight, Peirson received a signal from Hale, who was flying at a higher altitude (10 000 feet to Peirson's 8000) that he had detected radiation.[60] Hale's Dakota was contaminated, having spent an estimated 20 minutes in the cloud midway between Port Hedland and Broome. Peirson had not expected to find such contamination.[61] That same cloud headed east across the continent, to be detected by Townsville-based aircraft over Queensland. Peirson maintained during questioning at the Royal Commission that the radioactivity detected in the air was not deposited in great quantity on the Western Australian coast, although he conceded that there would have been some.[62] When pressed on the fact that an invisible cloud of radioactive particles would inevitably deposit those particles at some point, Peirson said there could 'possibly have been deposition somewhere between there and the east coast, like Townsville or somewhere like that'.[63] Counsel Peter McClellan suggested to Peirson that the whole reason for his mission was to enable the test authorities to make a public statement that the Australian mainland had not been contaminated.[64] Peirson replied '[a]t that time, referring to 4 October, the day after the explosion, it was quite possible to make a statement that there had been no deposition on the Australian mainland'.[65] He rejected the suggestion that these flights were done only for the sake of appearances.[66]

The Peirson mission out of Onslow operated separately to that of Hubert Gale in Broome, and the two met only briefly. They did not exchange any information, even though they were working some of the same sections of the coastline after the explosion. This lack of co-ordination seems extraordinary now, but was indicative of the strange way the air aspects of Hurricane were run. Former RAAF navigator William Bovill gave dramatic evidence during the Royal Commission in 1984. He was posted to Broome on 21 September 1952, under Wing Commander Roberts. He was a navigator in a Lincoln that had cannisters attached to sample the cloud, although he did not know what they were for when he first saw them. His job was to locate the Hurricane cloud and fly through it, at about 25 000 feet. The Lincoln was not pressurised, so the crew wore oxygen masks at that height. They were clad in ordinary air force flying suits.[67] Bovill and colleagues took off just after 10.30 pm on the night of the test from Broome, and flew through the night. They did not think they had located the cloud until they landed and were told that they had flown right through it. The instrument that was supposed to let them know they were in the cloud had not worked.[68] Bovill gave evidence that, until they landed, the crew thought that they had failed in their mission.[69] However, the cannisters were inspected by Gale and found to be 'red hot' with radioactivity.[70] After taking showers, the crew put on the same flying suits they had worn on the mission. McClellan asked, 'Was any question raised about whether or not you were contaminated?' Bovill replied, 'We knew we were contaminated'. McClellan continued: 'Did you express concern to the officers that you were contaminated?' Bovill answered simply, 'No sir'.[71]

The unorthodox arrangements for Hurricane air operations meant the usual lines of command were not always clear. The commander of 86 (T) Wing during Operation Hurricane was Wing Commander Robin Gray. Gray was forced out of the RAAF only months after Hurricane. Just before his sudden departure, he provided a letter to Headquarters, Eastern Area, Penrith, setting out the role of the 86 (T) Wing detachment of Dakota aircraft at Monte Bello. He noted that there was a skirmish between himself and the base commander at Onslow about who was commanding the Dakota crew. He won that battle but noted 'it cannot be over emphasised that for any future operation the detachment commander must maintain control of his aircraft as was done on this occasion'.[72] He described refuelling facilities as 'primitive', potentially limiting flying operations. He also said that conditions generally for crew were unsatisfactory: 'This was due to the normal difficulties associated with a tented camp in a remote area and the fact that the impending arrival of certain personnel was a closely guarded secret'.[73] (Just a few weeks after this report, Gray was allowed to resign his commission rather than face court martial for 'disgraceful conduct of an indecent kind' during a drunken night with junior RAAF personnel in Canberra.)[74]

Overall air support for Operation Mosaic was more extensive than it was for Hurricane, given the greater complexity of the operation, although the financial and personnel contribution of the RAAF was less than that for Hurricane. An RAF task force was formed under the command of Group Captain Paddy Menaul. Menaul had been attached to the US Nevada atomic test range, not long before his deployment to the Mosaic tests, and would later become air task commander for Operation Buffalo at Maralinga. The operation included seven Canberra bombers that would be used for cloud-sampling and photographic tasks. Both Canberra and Varsity aircraft were assigned to track the radioactive clouds. Varsities would also be needed for undertaking low-level radiological surveys and photography over the bomb craters. The Varsities were specifically charged with tracking fallout on the ground, and were equipped with radiological instruments that could measure this.[75] Menaul would also have command of two Whirlwind helicopters, operating out of Onslow, to assist with photographic duties and provide air communication between the mainland and the Monte Bello Islands.[76] These helicopters were difficult to maintain in the climatic conditions and were sometimes out of service due to complications from corrosion.[77] Nevertheless, a Whirlwind landed on Hermite Island two days after the arrival of *Narvik*, bearing mail and fruit, greatly boosting the morale of the troops and personnel not long arrived.[78] RAAF Neptunes were assigned to carry out safety patrols, watching for any shipping around the islands and helping to ensure that unauthorised craft were kept away. From mid-February, four Shackleton aircraft were tasked with daily weather observation flights over the Timor Sea.[79] The announcement specified that these flights would be watching out for tropical cyclones, given the local vernacular name 'willy willies' in the text.[80] In addition, four Hastings planes were allocated to maintain an internal transport service between Australian bases.[81]

In December 1955, William Penney wrote a curious and telling memo to Frederick Brundrett, chair of the Working Party on Operational Use of Atomic Weapons at the Ministry of Defence. He raised the issue of the air-sampling regimes proposed for Mosaic, as well as Operation Buffalo at Maralinga. He expected that the Australians would ask for filters flown through radioactive clouds by RAF planes. 'While I am not keen on the idea of giving them samples, I do not see how we can refuse.'[82] He indicated, somewhat bizarrely given the tight controls on air operations, that the Australians could just fly their own planes to get their own samples. Better yet, 'I am recommending that if they ask us we give them a little piece of the filters, but that we wait a few days so that some of the short-lived isotopes have decayed a good deal'.[83] That way, presumably, the extent of contamination would appear to be less.

While RAAF aircraft and crew took part, they were commanded by the RAF. The air operations base for Mosaic was RAAF Base Pearce, about 35 kilometres north of Perth. However, Pearce was not provided with facilities to deal with radioactive contamination. For that, RAAF Base Amberley, 40 kilometres south-west of Brisbane, was chosen.[84] RAAF air support consisted of transportation, signals, security patrols, search and rescue, and air traffic control.[85] For Mosaic G1, four RAF Canberra bombers flew from Pearce. Two of these conducted cloud sampling (around 20 minutes after firing) and two provided support. Both cloud-flying Canberras returned good samples to the radiochemical laboratories at Pearce.[86] The Canberras were monitored and decontaminated upon return to RAAF Pearce. Varsity aircraft surveyed the Onslow to Broome coastline one day after detonation.[87] After G2, the two Canberras again flew through the atomic cloud, and another Canberra followed behind, becoming 'moderately contaminated'.[88] The RAF contingent would later have to transfer to RAAF Base Edinburgh, near Adelaide, in the lead-up to the first Maralinga test series, Operation Buffalo.[89]

The risks were great for all aircrew in 1956, although perhaps less so than for Hurricane. An AWRE report on aircraft decontamination, prepared after Operation Mosaic and based on experiences from that test series, provided this background information about what the crew were flying into:

> Radioactive contamination after a nuclear explosion is primarily in the form of dust or smoke. After the fireball produced by the explosion cools the fission product material and other matter which composed the rest of the bomb, any supporting structures and sometimes a certain amount of earth, condense out as a dust cloud. The larger particles fall out fairly quickly but the finest particles remain airborne for some considerable period of time.[90]

While aircraft can collect any kind of dust, that from an atomic blast is reactive and chemically binds to the outer surfaces of the aircraft, making its removal difficult and hazardous. Radioactive contamination cannot be removed by burning or bleaching, so removing all traces of radioactivity was a severe challenge for the various air forces who sent crew and equipment.

The decontamination protocols for Mosaic were minimal, with a view to using the same aircraft for Operation Buffalo at Maralinga a few months later, in September and October 1956. Thorough decontamination was not undertaken, replaced simply by the use of AWRE barrier paint, which was sprayed onto the aircraft both before and after cloud-flying missions to seal in the radioactivity.[91] RAAF Base Pearce was not set up for radioactive decontamination, so the aircraft would not be properly processed until after Buffalo. An AWRE radiation scientist, David Stevenson, had invented the barrier paint for the AWRE, and was put in charge of managing aircraft contamination at Pearce. In his post-Mosaic report he said that it was 'anticipated that decontamination work on the Monte Bello Islands would be of a minor nature only'.[92] As the Royal Commission found, aircraft 'were sprayed initially and then re-sprayed after becoming contaminated, thus sealing the contamination and rendering it non-smearable'.[93] Stevenson said in his Royal Commission statement that this was the most sensible approach, reducing the exposure to the personnel as well as contamination of the Pearce airfield. 'By far the most radiation came from the engines, and after a few days decay the contribution from residual contamination on the outside of aircraft was negligible. The hazard was mainly one of ingestion and inhalation.'[94] Radioactive contamination on aircraft from nuclear weapons tests is highly adherent, so dealing with it requires protective clothing such as boiler suits with gloves, boots and caps. The evidence suggests, though, that these measures were not always taken. Stevenson noted that all contaminated materials and liquids from Pearce were transferred to Maralinga for recovery or disposal.[95] While Pearce was not to be used for decontamination, a personal change hut was eventually agreed to, where crew could change out of their contaminated clothing and have a warm, soapy shower. This 6-square-metre facility had a water heating system and wash basins, plus storage space. The hut was placed on a disused runway, near an 'active parking area' where the contaminated aircraft could be washed down. The washdown area was coated in a circular layer of sand mixed with oil, and effluent from the aircraft was drained to a 7570-litre tank.[96] This liquid, in total nearly 19 000 litres, was subsequently moved to Maralinga. Stevenson claimed that 'the activity was not greatly in excess of drinking water tolerance and all was allowed to soak away in a large pit'.[97] As well as the liquid waste, over 500 kilograms of solid waste was also sent from Pearce to Maralinga.[98]

Aircraft returning to Pearce contaminated would be allowed to sit in the active area for lengthy periods to enable the radioactivity to decay: 'no further attention was given to the aircraft for some days after the normal after-flight routine inspections had been completed'.[99] Refuelling a contaminated aeroplane was hazardous, potentially spreading the contamination to hoses and bowsers, and to personnel undertaking the task. 'Refuelling necessitates one man standing and walking over a highly contaminated wing. Overshoes were used to restrict the spread of contamination.'[100] Washing any contaminated clothing at Pearce was not possible. Clothing was an increasing problem, as Stevenson claimed in his statement; while trained scientific staff might have been able to keep relatively clean, inexperienced tradesmen 'tended to adhere to normal non-active routines and considerable contamination arose'.[101] The sand and oil circle proved to be 'an embarrassment'.[102] Much of the oil was washed out by rain and the top layer tended to blow around. Both the sand and oil collected on equipment, complicating efforts to contain the contamination. Despite the fact that Mosaic was the third British test series in Australia, some of the planning was remarkably deficient.

When word came through to *Narvik* that an RAAF aircraft had been found with radiation on the leading edge of its wings three days after G1, AWRE health controller James Hole was called upon to investigate. The aircraft, which Hole thought was a Hastings, had flown out of Onslow to RAAF Base Pearce, where the contamination had been detected. Hole, who was stationed aboard *Narvik*, gathered together 'a laundry basket full of all types of equipment' and was taken by helicopter from the islands to Onslow.[103] There he met up with ASIO officer Leo Carter, who had a Holden van. 'We found two or three bed mattresses which we laid out in the back of this Holden van. I laid down on the back of this van on my tummy', radiation measuring instruments by his side.[104] They carved the airstrip and adjacent areas into four quarters, then drove up and down each quarter slowly. He was able to detect radiation, which he described as patchy, and was also able to determine that the contamination had been caused by the G1 explosion.[105] This was evidence that contamination from Mosaic G1 had dropped onto the Australian mainland, despite assurances from authorities and politicians. This thirsty work took Hole and Carter into the only pub in Onslow, where Hole was happy to discover that 'there was no unusual activity above background' on the premises.[106]

Further, in questioning, Hole revealed that tests done at AWRE not long before Mosaic had shown that radiation detected by aircraft flying at 500 feet was about ten times lower than what was detected on the ground. Counsel to the Royal Commission Peter McClellan queried this, saying: '[y]ou knew before Mosaic that there was a difference between what the instrument in the aircraft would measure and the level of activity on the ground of about an order of 10?' Hole replied: 'That is right, providing the aircraft was flying at 500 feet, that was important'.[107] In effect, measurements taken in the air by surveys would have to be multiplied by ten to get an accurate measurement of the contamination on the ground. Hole maintained that this known issue was dealt with after G1 by flying a Varsity over Trimouille Island to calibrate the readings between the aircraft and the ground. However, this same calibration was not done for G2, owing to time pressure.[108] The absence of fixed monitoring stations on the ground meant that finding contamination on the mainland was effectively hit and miss. When the president of the Royal Commission, James McClelland, asked Hole 'after the experience of G1 when you realised that such a thing was possible [contamination being found on the mainland] did you take precautions in relation to G2 by putting monitoring stations on the ground', Hole answered with a direct 'No'.[109]

Aircraft did not have to be in the air to be contaminated. An RAF Canberra aircraft, on duty after the detonation of G2, had put in for refuelling at Port Hedland, where it was subjected to fallout simply by sitting on the tarmac overnight. The plane returned to Pearce with all upward-facing surfaces contaminated. Further investigation showed that the plane had been exposed to contaminated 'salty spray' that had fallen over the Port Hedland area the night after the test.

David Stevenson also gave evidence that fallout from Mosaic G1 had dropped on Onslow. His testimony was covered in the media, which reported that two DC3 aircraft had radioactive material spread across their upper surfaces.[110] While he did not consider the contamination to be too serious, he conceded that it would have been 'hazardous to anyone servicing the aircraft and picking up particles on their hands and feet'.[111] He later discovered that eight aircraft at Onslow had been contaminated. Onslow contamination was confirmed in a 1957 AWRE report, which found that '[m]easurable but minute contamination was ... observed along the coastal strip south and east of the Monte Bellos, in particular around Onslow'.[112] Air operations found radioactivity that could not be found in any other way, and were essential for providing glimpses of the toxins unleashed over Australia.

EPILOGUE

TIME AND TIDE

> It is not an exaggeration to say that what we learn from these trials may make all the difference between safety on the one hand and death or horrible suffering on the other for countless civilians if an atomic bomb attack should ever be launched against us.[1]
>
> AUSTRALIAN MINISTER FOR SUPPLY, HOWARD BEALE, 'WHY WE HOLD A-TESTS IN AUSTRALIA', 1956

> I have yet to find anyone who had any regrets at leaving.[2]
>
> REPORT OF PROCEEDINGS FOR HMS *ALERT*, 1956

Monte Bello is a closed book to most Australians. To a large extent, the British atomic tests at the nuclear archipelago are simply invisible, out of sight, out of mind. At the time, the bland assurances from two governments, notwithstanding deep (if fleeting) concerns in the Australian government about Mosaic G2, kept community anxiety at bay, and it quickly dissipated. However, quietly but with increasing pace, the events at Monte Bello have been undergoing investigation, historically, archaeologically, scientifically, artistically. While the British government still insists on retaining the central records of the test program, for no clear reason, it has re-released some after a mass withdrawal of files from the National Archives of the UK in 2018. The story of what went on will never be completely known so long as the British government chooses to remain stubbornly secretive. Secrecy always gives rise to suspicion, a lesson that the UK government has not heeded.

The Australian government was aware during the 1950s of concerns about ongoing radioactivity. A proclamation in the *Commonwealth of*

Australia Gazette of 4 April 1957 (No. 20) caused a small flurry of media interest. The proclamation cancelled the declaration of the Monte Bello Islands as a prohibited area, and declared another, smaller part of the islands prohibited. The minister for supply, Howard Beale, told the press that this was, 'in effect, the wiping out of the old large prohibited area proclaimed in connection with the Monte Bello tests, and the substitution for it of a much smaller area within the old one. The new area includes the islands themselves where the actual [blasts] took place'.[3] The next day, the *West Australian* stated that 'the danger is now calculated to be confined to the islands themselves and their territorial waters'.[4]

A news story in the *Sunday Telegraph* in August of 1958 was brief but indicated ongoing press interest in the aftermath of the Monte Bello tests. The story quoted the new minister for supply, Athol Townley, as saying the site was still radioactive. 'Mr. Townley said security precautions were taken to keep people from the areas. Monte Bello was still a prohibited area. Notices were placed prominently on every possible approach from the sea warning that the island was a radioactive area and dangerous to health.'[5] Although the story does not have a byline, government documents indicated that it was written by journalist Alan Reid, who had approached Townley with two questions, two days before publication. The story faithfully reported the statement issued by Townley.

The mystery of the 'radioactive Land Rover' is another enduring puzzle and a reminder of the aftermath of the tests. The vehicle had been used during Operation Hurricane and became highly contaminated, registering 2000 counts per second, about 20 roentgens per hour, which was extremely radioactive.[6] It had been brought to HMAS *Leeuwin*, a RAN naval base at Fremantle, in November 1953, along with a second contaminated vehicle used for surveying in 1953.[7] On 17 May 1960, the head of the Commonwealth X-Ray and Radium Laboratory, Donald Stevens – a former member of the AWTSC – had been asked to comment on the risk posed to the driver of the vehicle, an army corporal called Ronan, who said that he had driven it from Fremantle to Perth. Once there, according to Stevens, quoting information from the Department of Supply, unsuccessful attempts were made to decontaminate the vehicle, which only spread the contamination further. The navy then retrieved it and had a more successful go at decontamination.

While the vehicle used for surveying was decontaminated and sent to Sydney, the whereabouts of the highly contaminated vehicle remains a mystery. The rumour persists it was buried at HMAS *Leeuwin*, although the use of ground-penetrating radar failed to find it.[8]

The AWRE and its British civilian associates, the ministries of Defence and Supply, did some investigations of the aftermath of the Monte Bello tests, and as always, absolved themselves of responsibility for any harmful behaviour or aftermath. The Australians were left to find out for sure what the damage was. An agency of the Australian government, the Australian Radiation Laboratory (later the Australian Radiation Protection and Nuclear Safety Agency, ARPANSA), carried out some early surveys of environmental radiation in 1962 and 1968, and these formed a baseline against which the more detailed surveys in the 1970s and 1980s were measured.[9]

Both the Western Australian and the Australian governments became concerned about the site in the 1970s because of that age-old Australian pastime, fishing. Word had got out about the good fishing at Monte Bello, and people were making the ocean crossing to see if the stories were true. As the mining industry grew in that part of Western Australia, and with the port of Dampier and the new mining town of Karratha (established in 1968) expanding by the day, without recreational facilities matching the pace, many fit young men, in particular, were bored on the mainland and seeking the outdoors thrill of a trip to the islands and an impressive catch of fish.

Adding to concerns, a film crew had visited in the early 1970s and came back with a piece of steel imbued with cobalt-60, a highly radioactive isotope.[10] The crew also reported that a large amount of debris from the tests remained, in the form of structures and test objects, along with a distinct lack of vegetation in the ground zero areas. Pressure was mounting for a proper investigation and the era of surveys was about to begin. The Western Australian State X-Ray Laboratory decided to send an expert party to visit the islands to determine whether any hazard to the public remained. The naval officer-in-charge of the Western Australian Area was contacted and agreed to assist in transporting State X-Ray Laboratory personnel to the islands. The navy had a vessel, HMAS *Attack*, and patrol boats in the area, so this was easily arranged.

Two representatives from the Radiation Health Branch of the WA Health Department, BE King and LM Davies, confined their visit to Trimouille Island in October 1974, since the weather did not permit a landing at Alpha. HMAS *Attack* anchored off the northern end of the island in rough weather and the two scientists landed on a beach on the eastern side. They immediately found twisted galvanised iron that looked, from a distance, like a 'bedstead'.[11] They detected some radiation in the area, and it got stronger as they approached the fenced-off ground zero. All around was evidence of the long-ago test, including mangled remnants of the tower that had held the Mosaic G1 device. This material was radioactive. 'Some pieces of steel were quite large and heavy, while others were small and portable. Outside and towards the beach were the remains of a large electric motor and a pulley, all showing evidence of radioactivity.'[12] About 2 kilometres from ground zero, they found the wreckage of aeroplanes. The place was a junkyard, and 'hot' with it. Their later analysis showed that the chief culprits were cobalt-60, caesium-137 and europium-152, all toxic materials that are not safe to leave in the open, especially with free-ranging fishers arriving regularly. They concluded that the caesium and europium were fission products created in the nuclear chain reaction when the plutonium core of the weapon was detonated, while the cobalt isotope had resulted from the activation of cobalt in steel.[13] While they did not test for strontium-90, they believed it was present as well. 'The radiation levels at the ground zero area of Trimouille Island are well above the maximum permissible dose rates for continuous exposure of members of the public and above the levels considered acceptable for continuous exposure of radiation workers.'[14] They recommended maintaining the existing fences, adding some more and placing warning notices in both English and Asian languages around ground zero.[15]

Western Australian scientists B Hartley and R Murrell from the State X-Ray Laboratory visited the Monte Bellos in August 1977 as concerns continued to grow. They travelled aboard HMAS *Acute* and camped on a beach on Trimouille Island. This time, Trimouille, Alpha and Hermite were all visited.[16] On Alpha Island the scientists found significant radioactivity all around ground zero, including ferrous metal near the blast centre and on the nearby beach. According to the report, a 'large object found on the western beach of this island had the appearance of a neutron source holder [apparatus designed to hold neutron-emitting material] as it was filled with paraffin wax and had several holes into which tubes could be inserted with a rotating mechanism on the end'.[17] This visit provided more detail about the mess left by the British, some of which was still dangerously radioactive. The scientists encountered casual visitors from Geraldton during their visit, highlighting the need to consider the risks associated with people coming to the islands.[18] 'Apart from the effect these people may have on the possibly very fragile ecology on the islands, they could be exposed to radiation levels which, whilst not particularly dangerous in the short term, should be avoided if possible.'[19] They recommended declaring the islands a national park, and establishing two-yearly visits to check that the signs and fences remained in good order.[20]

A third survey was undertaken in 1978. A group of army personnel, along with Dr Malcolm Cooper from the Australian Radiation Laboratory, representatives from the Western Australian Museum and the State X-Ray Laboratory (and a crew member listed as 'Spud Murphy') took a five-day trip to the islands.[21] This group found that 'the significantly radioactive areas are confined to the two bomb sites'.[22] They also visited Hermite Island to investigate further the possibility of radioactive materials in the area of the control centre, perhaps deliberately taken there for disposal.

> None was found but questions still need to be answered on this matter. Several drums of concrete near the landing area showed higher than background radiation levels but the screenings of bluemetal also showed similarly higher radioactivity ... There are various pits dug and drums filled with soil which suggest possible burial of radioactive material.[23]

The group felt that while anything dangerous should be removed, 'unless these items constitute a physical hazard due to impending collapse or some other catastrophe that they should be left as a record of the activities on the island'.[24] Fred Wells from the Museum was tasked with collecting marine life samples to test for radioactive content, while army personnel Ian Reevie and George Hammel were asked to determine what work was needed to remove scrap metal and bury radioactive debris. The survey team found that the two bomb sites were the most significantly radioactive areas. The army members recommended that on Trimouille Island the radioactive areas could be cleaned by removing the radioactive iron tracking material, and removing and burying the concrete base at ground zero.[25]

An ABC TV *Four Corners* current affairs program on 14 October 1978 that made claims about the radioactive contamination left at Monte Bello prompted a response from Cabinet. A Cabinet minute of 17 October directed the Australian Ionising Radiation Advisory Council (AIRAC) to survey the islands.[26] The *Four Corners* program featured one of the dogged journalists who had covered Operation Hurricane from the mainland lookout, Dan O'Sullivan. He described a memorable sight for the program's audience, watching 26 years after the blast, many of whom would not have heard of the Monte Bellos or what happened there. 'Of course what we were viewing was … as awesome as watching a volcano explode … This sudden flash and this huge pillar of smoke and the knowledge that here was a cloud that was at its base about a mile across and that would be an area of total destruction, and beyond that would be the area of death.'[27] All those years later, one of the Nick's Nob gang still had a vivid picture in his head.

AIRAC surveyed the islands in 1978, as well as conducting surveys at the other two Australian test sites, Emu Field and Maralinga. A clean-up by the Royal Australian Engineers on behalf of the Department of National Development was carried out in 1979.[28] Called Operation Capelin, this program removed radioactive debris and soil from the ground-zero areas, and replaced it with fresh soil.[29] The various AIRAC reports, but particularly report no. 9, tabled in federal parliament in 1983, were blasted in the 1984 Kerr Report and later in the McClelland Royal Commission for being whitewashes. Until the Royal Commission, the AIRAC 9 report was considered the most comprehensive public account of the tests in Australia, going beyond the fairly basic site surveys carried out by the state and federal agencies during the 1970s to examine what the British actually did during the tests. AIRAC was essentially a reconstituted AWTSC, and therefore, arguably, too close to the events it was supposed to be objectively examining. Indeed, the Royal Commission report made the claim that 'at least some of the members of AIRAC did not approach their investigations with an open mind'.[30] Since the AWTSC did not place any serious impediments in the way of the tests during the 1950s, its successor organisation might be seen to be loath to cast doubt upon past internal decisions.

Much of the criticism from the Royal Commission was reserved for AIRAC's dismissal of severe problems associated with the tests at Emu Field, particularly the 'black mist', and numerous issues associated with Maralinga. The Royal Commission also found that the AIRAC report, without sufficient evidence, had absolved the British of any wrongdoing during any of its tests in Australia, including those at Monte Bello. Before the Royal Commission began, the Kerr Report had done a thorough demolition job on AIRAC 9. While the Royal Commission didn't like the 'extravagant language' Kerr had used in places to criticise AIRAC 9, the criticisms were still valid, and some were accepted by a member of AIRAC, Dr GM Watson, under examination during the Royal Commission.[31] The Kerr Report challenged multiple aspects of AIRAC 9, including 'fallout levels, progress of radioactive clouds, dosage estimates, risks of exposure of certain groups and aspects of management arrangements during the tests'.[32] Claiming that AIRAC always put the best possible spin on all British test activities, the Kerr Report 'found many examples of obfuscating use of language especially in those parts of AIRAC 9 which dealt with matters of political and public sensitivity'.[33] AIRAC hit back with a scathing response by its chair, Emeritus Professor Alan Clark, to the minister for home affairs and environment, Barry Cohen. Clark made counter-claims about Kerr's own objectivity, saying that his report had used 'pejorative language' with an intention to discredit AIRAC.[34] This unedifying squabbling was one of several factors that led to the establishment of the Royal Commission, set up to make a concerted effort to discover exactly what had happened. Despite all the surveys and reports over the years, no one in any Australian government actually knew.

The minister for mines and energy in the Hawke Labor government, Peter Walsh, announced the Royal Commission in July 1984. The public enquiry began taking evidence in August that year and reported its findings in November 1985. The transcript available on the National Archives of Australia website exceeds 10 400 pages, not including witness statements and a multitude of associated documents. Justice James McClelland's Royal

Commission was a thorough, expensive and ultimately enlightening public inquiry, even if it did not uncover the entire truth about the mostly secret program of British tests in Australia. What it did apart from anything else was open the door to thousands of hitherto hidden British files, starting a process of revealing the facts.

The human toll of the tests was substantial. Over and over, in testimony from those who witnessed the tests, claims were made that participants were not told about the risks, they were not provided with protective equipment when likely to be exposed to radioactivity, their health was not monitored when they returned home, and strange, life-long health problems arose for them and their families. The risks posed to all people in proximity to the Monte Bello tests were profound, but they only really started coming out into the wider public through, firstly, the Royal Commission in the mid-1980s, and later scholarly and media investigation.

The Royal Commission took the major step of moving its proceedings to London for three months, some six months after it began, in January 1985. This was done to apply more pressure on the Brits and to speak directly to key players such as William Penney, who was by then elderly and unwell, and clearly struggling with his memory. The hearings had been sufficiently lively and revelatory thus far to spark some dismissive public comments from the British. The new high commissioner to Australia, Sir John Leahy, was quoted a few months before the Royal Commission travelled to London saying that Britain's name was being 'dragged along the floor' by the evidence to date.[35] President of the Royal Commission Jim McClelland responded in characteristically robust fashion, saying, 'Sir John must have lived a very sheltered life or he would have noticed that even more unpleasant things have been said in the British papers about the British conduct of the nuclear tests'.[36]

Initially, the British government stone-walled the Royal Commission, going slow on the release of documents and generally showing little outward interest in proceedings. However, a stirring address by Justice McClelland upon the commencement of proceedings in London reached the intended target – Prime Minister Margaret Thatcher. Only Thatcher could push the Whitehall bureaucracy to act on the Royal Commission's requests, and she did. A huge media contingent at the first day of the London hearing heard McClelland say that it was 'only in recent weeks that the British Government has decided to be represented before the Commission. There are grounds for believing that this decision was taken reluctantly and only after the Commission had publicly suggested that the British Government was dragging its feet'.[37] This address to the British government from the bench came amid growing frustration that the British were being uncooperative and dismissive. McClelland said that continually invoking the need for secrecy to protect the national interest had

> always been a convenient alibi for failure of disclosure. But today it is hard to believe that Britain is in possession of any atomic secrets unknown to the great Nuclear Powers. We are not here to poke our noses into British technical secrets, but there is a certain minimum of information to which, as the host country to the nuclear tests, we feel entitled to have access.[38]

Soon, thousands of documents descended on the Commission, placing a huge strain on the legal teams, who had to process complex information quickly to ensure that they got the most out of their examination of witnesses during their limited time in the British capital. The apparent arrogance in first withholding these documents, and then flooding the Commission with them, was not appreciated by those who had to deal with it.

Just before the Royal Commission got underway, and in the face of rising concern among British members of parliament, Margaret Thatcher ordered the first epidemiological study of the health of nuclear veterans, which began in 1985. The study of more than 22 000 veterans was carried out by the National Radiological Protection Board (NRPB). This study, and two subsequent ones, reported only small risks of developing leukaemia and multiple myeloma, but claimed that nuclear veterans were in general healthier than the broader population.[39] These studies have long been despised by the nuclear veterans community, which has campaigned for recognition of their ongoing suffering and the genetic consequences for their families. One organisation representing nuclear veterans, Labrats, quoted a forerunner group, the British Nuclear Test Veterans Association, as saying in the 1980s that 'this study was not fit for purpose, the methods were questionable, and the outcomes were just a statistical exercise and proved nothing'.[40] Those views have not changed among UK nuclear veterans, as evidenced by recent Labrats campaigning. The appointment of the NRPB to undertake the study was criticised robustly at the time, including by Liberal member David Alton, later Baron Alton of Liverpool, who said on 12 March 1984 that it was a case of 'the watchdog being too closely identified with the burglar'.[41] The NRPB had expertise in monitoring of radiation exposure, but not in surveying the health of those affected. In response to Alton's concerns, the minister of state for defence procurement in Thatcher's government, Geoffrey Pattie, commented, 'I hope that [Mr Alton] can maintain the appropriate objectivity on this matter. It is a difficult subject that is prone, understandably, to great emotion and to press reporting that is long on sensationalism and short on facts and analysis'.[42] Pattie said without equivocation that the nuclear veterans were not exposed to harmful radiation, the official line that the British government has run consistently. A 2022 review of four decades of nuclear test veteran epidemiology, including recent data up to 2017, did find higher rates of certain types of leukaemia, as well as some solid cancers such as stomach, bladder and prostate, and certain respiratory cancers.[43] However, the authors cautioned that the data they analysed did not support a clear, causal link between test-related radiation exposure and widespread excess cancer or mortality among UK nuclear veterans. They called for continued follow-up to resolve remaining uncertainties.[44]

Thatcher's government was not pleased with the outcome of the Royal Commission. Cabinet minutes indicate the view that the resulting report 'seemed to be greatly coloured by the benefit of hindsight enjoyed by those who had written it'.[45] That Cabinet meeting made the point that it would be necessary 'to counter any public tendency to jump to the conclusion that the report was right in calling upon the United Kingdom to provide £100 million for the cleansing of the three former test sites'.[46] Members were mindful that 'any person who had been near the sites during the period of British nuclear tests in Australia would be likely to try and blame the British Government for any disease that they contracted'.[47] Those present agreed that the official public position was that Thatcher and her colleagues were studying the report carefully. Briefing notes for Thatcher ahead of a visit by Prime Minister Bob Hawke in 1986 stressed the importance of no quarter being given on the question of contributing to any clean-up. The notes state that the UK government was not to let Hawke 'draw you down the path of offering to contribute to further clean-up of the test site. We are ready to give technical help but there is no justification for expecting us to bear any further clean up costs'.[48] The Brits continued that stance for some years, until finally forced to pay for a portion of clean-up costs at Maralinga in 1993 in light of new information about contamination there. However, the UK government has never accepted responsibility for the radioactive contamination left in Australia, and it never attempted to clean up the Monte Bello Islands.

On 8 July 1992, the federal minister for primary industries and energy, Simon Crean, stood on Trimouille Island and officially handed the islands over to the Western Australian government.[49] The handover from the federal to the state government, officially gazetted the day before, was the culmination of 15 years of negotiation.[50] Placed under the Department of Conservation and Land Management, the islands became a conservation park.[51] The federal government contributed $300 000 to help manage the islands, including work to eradicate rats and cats. The islands were safe for visitors, said the state environment minister, Bob Pearce, although 'part of Trimouille island and the actual test sites – which would be maintained as historical sites – would be restricted to day-visits only'.[52] That guidance has since been revised to one-hour visits to the two ground zero sites.

The distress of the veterans and civilians, mostly from the UK but also from Fiji and New Zealand, was greatly increased in 2012 when the UK Supreme Court refused (by a slim majority) to allow a case for compensation against the British Ministry of Defence to go forward.[53] A group of 1011 veterans and civilians had launched the action, but it came to nothing. The grounds for refusing the case were that the events in question fell outside the scope of the *Limitations Act 1980*, under which compensation claims must be made no more than three years after the issues became apparent.[54] Legal options are not necessarily exhausted, however. At present, potential litigation against the Ministry of Defence for the release of medical records withheld from veterans is being contemplated as part of a broad campaign seeking accountability. This campaign, led by nuclear veterans, their families and allies, has consulted with the London-based law firm McCue Jury & Partners. In addition to a possible civil claim, a 500-page criminal dossier of evidence was provided to the Metropolitan Police in May 2025.[55] The British government is reviewing allegations that medical records, including blood and urine test results obtained between 1952 and 1967, were deliberately withheld from its own military personnel. The UK minister for veterans and people, Al Carns, made a statement in the House of Commons in June 2025 that tens of thousands of relevant policy documents and records were under review, although no completion date was given.[56] A tranche of documents known as the 'Merlin database', which contains over 28 000 records of historic technical and scientific documentation on the UK's nuclear testing program, is being transferred to the National Archives, following sustained public pressure from veterans and their allies.

The efforts of the nuclear veterans have only been belatedly acknowledged more broadly by the British political class, and at time of writing cracks in the Ministry of Defence bulwarks were appearing. On 28 November 2023, Labour member Rebecca Long Bailey told the House of Commons that the NRPB report 'was only seven pages long, and it was criticised by veterans as a whitewash. It was repeated, with similar results, in 1993 and 2003. Further, it has been reported in recent days that the 1988 scientific report was altered on the instruction of officials'.[57] Long Bailey asserted that the nuclear test veterans had suffered severe physical and psychological consequences from their service, saying that they had 'the same rate of DNA damage as clean-up workers at Chernobyl'.[58] Their records of exposure were often either absent altogether or withheld from them.

Britain has never properly reckoned with its nuclear test activities. To this day, secrecy reigns. As British journalist Susie Boniface poignantly put it in her 2024 book focusing on the plight of British veterans and others: when asked 'who's to blame', she said that there was 'only one answer: whoever is currently Prime Minister. They alone have the power to end the longest-running scandal in British history'.[59] Not one British prime minister has fully answered the call of their own country's nuclear veterans for acknowledgement of the risks and harms they faced, despite some political support for their cause and the creation of a ceremonial medal for nuclear veterans in 2022.[60] At the time of the Australian Royal Commission, some

British politicians called for more British accountability, including David Alton, who said in the House of Commons on 19 December 1984: 'It is ironic that we should have the same Queen but that different ground rules should apply. It is a paradox that the Australians should have established a Royal Commission but that what is good enough in Australia is not good enough here'.[61] Just a couple of weeks later, the Royal Commission landed in London and grabbed what it could of the records to share with the Australian public. In doing so, it unleashed Australian anger towards the British government that had rarely been seen. The Australian media picked up every admission and revelation from those dangerous times and ran headline stories for 18 months.

The scale of the environmental effects of the tests is still being uncovered. Recent scientific investigation of the aftermath of both Hurricane and Mosaic has found evidence of ongoing marine sediment contamination at the islands. A 2019 study discovered plutonium in sea turtles nesting in the islands' contaminated sands.[62] The researchers made the claim that the higher concentrations in the decade immediately after the tests in the 1950s 'may have resulted in lethality or health impacts to a generation of turtles that likely carry biomarkers today'.[63] Indeed, the Hurricane test took place during the nesting season for green, hawksbill and flatback sea turtles, in October.[64] Turtles can live for 60 years or more, and their exposure to the fallout from the Monte Bello bombs would have had long-lasting effects on breeding populations in the area, including possible genetic damage. The researchers compared Monte Bello to the Pacific nuclear test site Bikini Atoll and the Japanese nuclear accident site Fukushima as areas of severe contamination of the natural environment.[65]

Another study by some of the same researchers found that while short-lived radionuclides such as cobalt-60 had decayed, long-lived plutonium isotopes were present in island soils, dunes and beaches as well as in the terrestrial and marine biota that were tested. Levels of radioactivity were still elevated and would remain so for tens of thousands of years.[66] Each of the Monte Bello tests left a distinct 'signature' that could be seen in sediments and animals. Erosion was steadily moving radioactive particles from the land to the sea, so while some types of radiation near the ground zeroes had decreased, the plutonium and other long-lived radionuclides are still a management concern. Wildlife such as reptiles, mammals, fish, oysters and sea cucumbers showed detectable but low radionuclide uptake.[67]

While the islands themselves have been scientifically surveyed over the years, with a focus on the radioactive contamination of the sands and soils, the marine environment surrounding them had not been until recently. A study published in 2024 showed that marine sediment around the islands remains high in plutonium isotopes, along with americium-241, a radioactive decay product of plutonium-241. The highest concentrations are to be found in sediment adjacent to the site of the massive Mosaic G2 detonation on Alpha Island.[68] The levels recorded were several orders of magnitude above normal background radiation in the area.[69] The Monte Bello Islands carry a long-lasting radioactive legacy, including the highly dangerous isotope plutonium-239, the material at the core of all three weapons tested there, which has a half-life of 24 000 years. As a recently published paper on the residual radioactive contamination at Monte Bello put it, the 'Marine Park's unique mixture of residual radionuclides in high value ecological habitats and recreation areas ... requires ongoing management'.[70] This study represents the opening of a new chapter in investigating the radioactive legacy left behind at the archipelago by nuclear tests that were conducted with little regard for the natural environment.

Today, visitors are advised not to spend more than one hour at the ground zeroes of the two Mosaic bombs. The risk is real and the one-hour time limit based on sound science. At G2 on Alpha Island, walking to the top of a rise near a concrete plinth marking ground zero reveals the shape of a crater that was formed when a fundamental force of the universe was unleashed and the tranquil islands resounded with the mighty boom of the largest explosion ever on Australian soil. Unlike at the test sites at Maralinga and Emu Field, there is no strange, radioactive glass formed when sand melted from the intense heat. Because the terrain is predominantly limestone, this telltale feature of atomic testing is absent. Instead, the bomb authorities found the crater and surrounding areas coated in a 'soft chalky powder'.[71] The environment cannot and will not ever be the same, but radiation is invisible, so the casual visitor may not fully appreciate what happened here. There are signs in several places, although not many, carrying warnings. Old signs no longer standing but left lying on the ground warn people in six or seven different languages, but the newer signs are in English only. A plaque placed by a veterans' association at the G1 site wrongly identifies it as G2, perhaps indicating that the group that erected it had confused Trimouille and Alpha islands. The sign is touching all the same, since it hints at the world of pain inflicted on the nuclear veterans who have lived with the physical and psychological aftermath of the Monte Bello tests for decades. These islands, so far away from the country that chose them for its most dangerous tests, well away from most of its own citizens, were unknown and unknowable, and Britain would have liked to have kept it that way. That country walked away from its responsibilities, and continues to maintain that it did nothing wrong. That position is increasingly hard to sustain, however, especially when the history is viewed in all its aspects.

The Monte Bello tests provide a glimpse into the political machinations and scientific advances that were occurring in Cabinet offices and scientific labs halfway around the world from Australia. The Australian government and people had little input into either of those closed domains, and the imposition of strict secrecy meant that these activities were little discussed, other than in superficial ways, in this country. And yet, the events at Monte Bello concerned all Australians. Fallout from all three tests travelled across the continent, and left their immediate environments tainted. First Hurricane, then later Mosaic, added their unique inscriptions to a palimpsest on those ancient, beautiful islands whose layers are only now being discerned. Australia gave away its territory too cheaply back in the days when it still felt the tight ties of colonialism. By the time the British scientists and military personnel boarded their ships, turned their gaze wistfully homeward and departed the Monte Bello Islands, it was too late. The damage had been done.

APPENDIX

BRITISH ATOMIC TESTS IN AUSTRALIA

MAJOR TRIALS

Operation Hurricane
Monte Bello Islands, Western Australia
3 October 1952, 25 kilotons

Operation Totem
Emu Field, South Australia
Totem 1: 15 October 1953, 9.1 kilotons
Totem 2: 27 October 1953, 7.1 kilotons

Operation Mosaic
Monte Bello Islands, Western Australia
Mosaic G1: 16 May 1956, 15 kilotons
Mosaic G2: 19 June 1956, between 60 and 98 kilotons (disputed)

Operation Buffalo
Maralinga, South Australia
Buffalo 1 (One Tree): 27 September 1956, 15 kilotons
Buffalo 2 (Marcoo): 4 October 1956, 1.5 kilotons
Buffalo 3 (Kite): 11 October 1956, 3 kilotons
Buffalo 4 (Breakaway): 22 October 1956, 10 kilotons

Operation Antler

Maralinga, South Australia

Antler 1 (Tadje): 14 September 1957, 1 kiloton
Antler 2 (Biak): 25 September 1957, 6 kilotons
Antler 3 (Taranaki): 9 October 1957, 25 kilotons

MINOR TRIALS

Kittens

Emu Field, South Australia

September–October 1953

Maralinga, South Australia

(Naya): May–June 1955
(Naya): March 1956
(Naya): March–July 1957
(Naya): March–July 1959
(Naya): May 1961

Tims

Maralinga, South Australia

(Naya) July 1955
(Kuli/Naya): March–July 1957
(Kuli): September–November 1957
(Kuli): April–June 1958
(Kuli): September–November 1958
(Kuli): May–November 1959
(Kuli): April–October 1960
(Kuli/Naya): August 1961
(Kuli): March–April 1963

Rats
Maralinga, South Australia
(Naya): April–June 1958
(Naya): September–November 1958
(Dobo): March–July 1959
(Naya and Dobo): September 1960

Vixen A
Maralinga (Wewak), South Australia
June–August 1959
May–August 1960
March–April 1961

Vixen B
Maralinga (Taranaki), South Australia
September–October 1960
April–May 1961
March–April 1963

Adapted from the Royal Commission into British Nuclear Tests in Australia, *Report* (JR McClelland, president), vol. 1, Appendix G, Chronology, pp. vii–1 to vii–10.

GLOSSARY

Alpha particles Positively charged particles containing two protons and two neutrons that are emitted by certain radioisotopes, particularly those with a high atomic number.

Alpha radiation Radiation caused by alpha particles. Alpha radiation has very little penetrating power but may present a serious hazard if alpha particles are inhaled or ingested.

Atom The smallest particle of an element that retains the characteristics of that element. It is made up of a nucleus and a cloud of surrounding electrons.

Base surge The large upward movement of radioactive water into the air induced by a nuclear explosion in water.

D-notice A secret government request to senior media representatives not to publish certain specified details about defence- or security-related activities. The D-notice system was adopted in Australia in 1952.

Dose The amount of energy delivered to a mass of material by ionising radiation passing through it.

Dosimeter A device, instrument or system used to measure or evaluate a dose of radiation.

Fallout The descent to the Earth's surface of particles contaminated with radioactivity, following the dispersion of radioactive material into the atmosphere by nuclear explosion. The term is applied both to the process and, in a collective sense, to the particulate matter.

Film badge A plastic holder containing a piece of film similar to a dental x-ray film and worn by personnel at a nuclear test. Radiation passes through the paper and exposes the film. The film is developed and the degree of darkening is a measure of the radiation dose received. The film holder usually contained metal filters to enable discrimination between different types of radiation.

Fission The process in which the nucleus of a heavy element such as uranium or plutonium splits into two nuclei of lighter elements, accompanied by the release of substantial amounts of energy.

Fusion The process in which the nuclei of light elements such as hydrogen, deuterium or tritium combine to form the nucleus of a heavier element, accompanied by the release of substantial amounts of energy.

Gamma radiation Penetrating electromagnetic radiation emitted from the nucleus of radioactive elements. This form of radiation is most readily measured by monitoring equipment such as film badges and dosimeters.

Half-life The time in which the activity of a radioactive species will decline to half its initial value by radioactive decay. The half-life of a radioactive species is a characteristic property of that species.

Health physics The science of human health and radiation exposure – a branch of medical science devoted to radiation safety.

Ionising radiation Radiation that integrates with matter to add or remove electrons from the atoms of the material absorbing it, producing electrically charged (positive or negative) atoms called ions.

Isotopes Forms of the same element whose nuclei contain different numbers of neutrons and therefore have different mass numbers. Isotopes of an element have nearly identical chemical properties but differ in their nuclear properties. For instance, some isotopes of an element may be radioactive, and others not.

Major trials Atomic tests conducted at Monte Bello Islands, Emu Field and Maralinga in Australia that involved detonating a complete atomic bomb, resulting in a 'mushroom cloud'.

Minor trials Hundreds of tests conducted at Emu Field and Maralinga in Australia that involved examining how radioactive materials and atomic weaponry would behave under various conditions such as fire or conventional explosion.

Neutron A nuclear particle with no electric charge and a mass approximately equal to or slightly greater than that of a proton. Neutrons are present in all atoms except those of the lightest isotope of hydrogen.

Nuclide Species of atoms having a specified number of protons and neutrons in their nuclei. Radionuclides are the radioactive forms of nuclides. They are often expressed as, for example, ^{239}Pu, which shows the number of neutrons and hence the form of isotope.

Plutonium (Pu) A dense, silvery radioactive element that does not occur naturally but is made in a reactor by bombarding uranium with neutrons. It was first produced in 1940. Plutonium has 13 known isotopes.

Radioactivity The property of certain radionuclides of spontaneously emitting particles and/or x-ray or gamma ray radiation, or of undergoing spontaneous fission. The rate of decay is specific to a given species of radionuclide and cannot be changed by known physical or chemical processes.

Radionuclide A radioactive nuclide. See Nuclide.

Thermonuclear weapon A nuclear device that relies on raising the temperature of a mixture of deuterium and tritium nuclei to above 10 million degrees Celsius, at which point nuclear fusion reactions occur. This type of weapon is also known as a hydrogen bomb or fusion bomb.

Warhead The explosive head of a bomb.

Yield The amount of energy generated by a nuclear explosion, usually expressed in kilotons (for fission devices) or megatons (for fusion devices). A kiloton is equivalent to 1000 tons of TNT and a megaton is equivalent to one million tons of TNT.

Adapted from a variety of sources, in particular the Royal Commission into British Nuclear Tests in Australia, *Report* (JR McClelland, president), vol. 2, pp. II–1 to II–9; JL Symonds, *A History of British Atomic Tests in Australia*, pp. 585–93; and the Australian Government Department of Education, Science and Training website: <www.radioactivewaste.gov.au/Glossary.htm>, accessed 6 May 2008.

NOTES

Prologue

1 Frank L Hill, 'Notes on the Natural History of the Monte Bello Islands', *Proceedings of the Linnean Society of London*, vol. 165, issue 2, June 1955, p. 113.

2 KD Morris, 'Management Proposals for the Montebello Islands and Surrounding Waters', CALM Occasional Paper No. 3/91, August 1991, Department of Conservation and Land Management, Como, Western Australia.

3 Peter Veth, Ken Aplin, Lynley Wallis, Tina Manne, Tim Pulsford, Elizabeth White and Alan Chappell, *The Archaeology of Montebello Islands, North-West Australia: Late Quaternary Foragers on an Arid Coastline*, Archaeopress, Oxford, 2007, p. i.

4 PF Perry and FE Wells (eds), 'Survey of the Marine Fauna of the Montebello Islands, Western Australia and Christmas Island, Indian Ocean', *Records of the Western Australian Museum*, Supplement no. 59, Western Australian Museum, November 2000, p. xi.

5 Hill, 'Notes on the Natural History of the Monte Bello Islands', p. 113.

6 Veth et al., *The Archaeology of Montebello Islands*, p. i.

7 Margaret Gowing, 'Monte Bello', in *Independence and Deterrence: Policy Execution*, vol. 2, Palgrave Macmillan, London, 1974, p. 484.

8 Leonard Tyte, Operation Hurricane: Dr Tyte's Diary, 1952. TNA: ES 1/78.

9 Captain George Colville, commanding HMS *Zeebrugge*, senior officer of the 4th Landing Ship Squadron, Report of Proceedings, Appendix II, 'Weather', 6 June 1952. NAA: A13070, SMITH 168.

10 PD Montague, 'The Monte Bello Islands', *The Geographical Journal*, July 1913, vol. 42, no. 1, p. 35.

11 I Ward, P Veth and T Manne, 'To the Islands Born: The Research Potential of Submerged Landscapes and Human Habitation Sites from the Islands of NW Australia', in J Harff, G Bailey and F Lüth (eds), *Geology and Archaeology: Submerged Landscapes of the Continental Shelf*, Geological Society, London, Special Publications, vol. 411, 2016, p. 255.

12 Ariana BJ Lambrides, Kane Ditchfield, Sean Ulm and Peter Veth, 'A Moving Feast: Australia's Coastal and Island Archaeological Record', in Scott M Fitzpatrick and Jon Erlandson (eds), *The Oxford Handbook of Island and Coastal Archaeology*.

13 Lambrides et al., 'A Moving Feast'.

14 Cassia J Piper and Peter M Veth, 'Palaeoecology and Sea Level Changes: Decline of Mammal Species Richness During Late Quaternary Island Formation in the Montebello Islands, North-Western Australia', *Palaeontologia Electronica*, vol. 24, no. 2:a20, 2021, p. 2.

15 Ward et al., 'To the Islands Born', p. 252.

16 Ward et al., 'To the Islands Born', p. 252.

17 Ward et al., 'To the Islands Born', p. 254.

18 Ward et al., 'To the Islands Born', p. 257.

19 Ward et al., 'To the Islands Born', p. 257.
20 Ward et al., 'To the Islands Born', p. 257.
21 Ward et al., 'To the Islands Born', p. 257.
22 Ross Anderson and Theresa Archer, 'Trial and Errors', *The Great Circle*, vol. 44, no. 2, 2022, pp. 67–73.
23 Trial (1622/05/24), Trial Rocks, Western Australian Shipwreck Databases, Western Australian Museum, <museum.wa.gov.au/maritime-archaeology-db/wrecks/trial>, accessed 9 January 2025.
24 Anderson and Archer, 'Trial and Errors', p. 68.
25 Anderson and Archer, 'Trial and Errors', p. 67.
26 Marsden Hordern, 'Australia's Debt to Stokes and the BEAGLE', *Maritime Studies*, vol. 1992, issue 64, p. 13.
27 Marsden Hordern, 'Darwin's Ship "Beagle" and Her Captains', *Journal of Australian Naval History*, vol. 7, no. 1, 2010, p. 26.
28 Montague, 'The Monte Bello Islands', p. 39.
29 Montague, 'The Monte Bello Islands', p. 40.
30 Montague, 'The Monte Bello Islands', p. 40.
31 Hill, 'Notes on the Natural History of the Monte Bello Islands', p. 114.
32 Myra Stanbury, 'Mother-of-pearl Shell Cultivation: An Early 20th Century Experiment in the Montebello Islands', *The Great Circle*, vol. 16, no. 2, 1994, p. 91.
33 Stanbury, 'Mother-of-pearl Shell Cultivation', p. 94.
34 Stanbury, 'Mother-of-pearl Shell Cultivation', p. 95.
35 Stanbury, 'Mother-of-pearl Shell Cultivation', p. 98.
36 Stanbury, 'Mother-of-pearl Shell Cultivation', p. 99.
37 Hill, 'Notes on the Natural History of the Monte Bello Islands', p. 115.
38 Hill, 'Notes on the Natural History of the Monte Bello Islands', p. 115.
39 Hill, 'Notes on the Natural History of the Monte Bello Islands', p. 115.
40 Hill, 'Notes on the Natural History of the Monte Bello Islands', pp. 119–23.
41 Margaret Gowing, 'Monte Bello', in *Independence and Deterrence: Policy Execution*, vol. 2, Palgrave Macmillan, London, 1974, p. 478.
42 John Moroney and Malcolm Cooper, Australian Radiation Laboratory, Environment Radiation at the Monte Bello Islands from Nuclear Weapons Tests Conducted in 1952 and 1956, December 1982. NAA: A6456, R013/004.
43 Mark S Bell, 'Beyond Emboldenment: How Acquiring Nuclear Weapons Can Change Foreign Policy', *International Security*, vol. 40, no. 1, Summer 2015, p. 101.
44 Bell, 'Beyond Emboldenment', p. 104.
45 Duncan Sandys, Defence: Outline of Future Policy (Defence White Paper), March 1957, TNA, CAB/129/86, p. 7.
46 Bell, 'Beyond Emboldenment', p. 107.
47 Harold Wilson, leader of the opposition, Debates, House of Commons Hansard, vol. 687, Thursday 16 January 1964.
48 John Baylis, 'The Development of Britain's Thermonuclear Capability 1954–61: Myth or Reality?', *Contemporary Record*, vol. 8, no. 1, Summer 1994, p. 171.
49 Richard Moore, 'British Nuclear Warhead Design 1958–66: How Much American Help?', *Defence Studies*, vol. 4, no. 2, 2004, p. 216.
50 Royal Commission into British Nuclear Tests in Australia, *Report* (JR McClelland, president), 2 vols, AGPS, Canberra, 1985, p. 122.
51 Dorrie Walley, oral history, Burrup Stories Project, March 2010. Creator: Kate Gregory. Australian Institute of Aboriginal and Torres Strait Islander Studies, GREGORY_KOO1.

Chapter 1: A maritime nation goes atomic

1 *Daily Express*, 15 October 1953, p. 4.
2 Frederick Lindemann, Note to Churchill about the British position on the atomic bomb, 19 January 1950. Lindemann/Cherwell papers J97, Nuffield Library, Oxford.
3 Lindemann, Note to Churchill, 19 January 1950.
4 Lindemann, Note to Churchill, 19 January 1950.
5 Peter G Boyle, 'Britain, America and the Transition from Economic to Military Assistance, 1948–51', *Journal of Contemporary History*, July 1987, vol. 22, no. 3, July 1987, p. 525.
6 Lindemann, Note to Churchill, 19 January 1950.
7 Margaret Gowing, 'Deterrence Recalculated', in *Independence and Deterrence: Policy Making*, vol. 1, Palgrave Macmillan, London, 1974, p. 222.
8 Norman Brook, Cabinet secretary, Clandestine Use of Atomic Weapons, Top-secret memorandum to Clement Attlee, prime minister, 12 July 1951. TNA: PREM 11/560.
9 Brook, Clandestine Use of Atomic Weapons, July 1951.
10 Draft press notice, 19 February 1951, annotated by Winston Churchill, CHUR 2/28, Churchill Archives Centre, University of Cambridge.
11 Draft press notice, February 1951.
12 Draft press notice, February 1951.
13 Lorna Arnold and Mark Smith, *Britain, Australia and the Bomb: The Nuclear Tests and Their Aftermath*, 2nd edn, Palgrave Macmillan, New York, 2006, p. 10.
14 Margaret Gowing, with Lorna Arnold, *Independence and Deterrence: Britain and Atomic Energy, 1945–1952*, United Kingdom Atomic Energy Authority, St Martin's Press, New York, 1974, p. 406.
15 Winston Churchill, Top-secret Cabinet document, 14 December 1954: 'Notes on Tube Alloys, 1954'. DSND 14/4 Churchill Archives Centre.
16 Churchill, 'Notes on Tube Alloys, 1954'.
17 'Man who Hates Fuss to Blast Atom Bomb', Sydney *Daily Mirror*, 15 July 1952, p. 10.
18 'The Scientific Civil Service: Reorganisation and Recruitment during the Reconstruction Period', *Aircraft Engineering*, October 1945, p. 280.
19 Lorna Arnold, 'Lord Penney', *Proceedings of the American Philosophical Society*, vol. 136, no. 3, September 1992, p. 445.
20 Frederick Lindemann, Note 1 to Churchill, 1 December 1953. Lindemann/Cherwell papers J97, Cherwell 1950, Nuffield College Library, Oxford.
21 Frederick Lindemann, Note 2 to Churchill, 1 December 1953. Lindemann/Cherwell papers J97, Cherwell 1950, Nuffield College Library, Oxford.
22 Lindemann, Note 2 to Churchill, 1 December 1953.
23 David Reynolds, 'A "Special Relationship"? America, Britain and the International Order Since the Second World War', *International Affairs*, vol. 62, no.1, Winter 1985–1986, p.11.
24 Tim Sherratt, 'A Political Inconvenience: Australian Scientists at the British Atomic Weapons Tests, 1952–53', *Historical Records of Australian Science*, vol. 6, no. 2, December 1985, p. 139.
25 Frederick Lindemann, Note to Churchill, 29 December 1952. Lindemann/Cherwell papers, J122, Nuffield College Library, Oxford.
26 JL Symonds, *A History of British Atomic Tests in Australia*, AGPS, Canberra, 1985, p. 8.
27 James Purdon, 'The Meaning of Monte Bello', in John Beck and Ryan Bishop, *Cold War Legacies: Legacy, Theory, Aesthetics*, Edinburgh University Press, Edinburgh, 2016, p. 85.
28 Purdon, 'The Meaning of Monte Bello', p. 85.

29 Lord Tedder, BJSM Washington, Top-secret cypher telegram to Ministry of Defence London, 27 July 1950. TNA: FCO 1/1.
30 Elliot, Ministry of Defence, London, Top-secret cypher telegram to Lord Tedder, BJSM Washington, 14 August 1950. TNA: FCO 1/1.
31 Lord Tedder, BJSM Washington, Top-secret cypher telegram to Ministry of Defence London, 17 October 1950. TNA: FCO 1/1.
32 Testing of United Kingdom Atomic Weapon, Top-secret ministerial minute paper prepared for the UK Prime Minister Clement Attlee, 14 September 1950. TNA: FCO 1/1.
33 Telegram from Commonwealth Relations Office Secretary London (draft), 20 September 1950. TNA: FCO 1/1.
34 CRO telegram, September 1950.
35 Clement Attlee, UK prime minister, telegram to Robert Menzies, Australian prime minister, 21 September 1950. TNA: FCO 1/1.
36 Robert Drewe, *Montebello: A Memoir*, Hamish Hamilton, Melbourne, 2012, p. 35.
37 Attlee letter to Menzies, 16 September 1950.
38 Attlee letter to Menzies, 16 September 1950.
39 Symonds, *A History of British Atomic Tests in Australia*, p. 18.
40 UK Services Liaison Staff Australia, Top-secret cypher telegram to Ministry of Defence London, 13 October 1950. TNA: FCO 1/1.
41 Symonds, *A History of British Atomic Tests in Australia*, p. 18.
42 Symonds, *A History of British Atomic Tests in Australia*, p. 18.
43 Howard Beale, *This Inch of Time: Memoirs of Politics and Diplomacy*, Melbourne University Press, Melbourne, 1977, p. 78.
44 Stewart Menaul, 'Great Britain and NATO Theater Nuclear Forces', *Strategic Review*, United States Strategic Institute, Washington DC, Winter 1981, p. 61.
45 Stewart Menaul, Evidence to the Royal Commission, Transcript of Proceedings. NAA: A6448, 12, p. 6984.
46 UK Services Liaison Staff Australia, Operation Hurricane timeline, 17 July 1952. TNA: FCO 1/4.
47 UKSLS, Operation Hurricane timeline, July 1952.
48 UKSLS, Operation Hurricane timeline, July 1952.
49 UKSLS, Operation Hurricane timeline, July 1952.

Chapter 2: First stirrings of a hurricane

1 Kevin Ruane, *Churchill and the Bomb in War and Cold War*, Bloomsbury Academic, London, 2016, p. 211.
2 Captain FB Lloyd, Ministry of Supply, London, Top-secret letter to Neil Pritchard, Commonwealth Relations Office, 7 December 1953, TNA: FCO 1/6.
3 Chapman Pincher, 'Atom Bomb Wizard', *Western Mail* (Perth), 6 March 1952, p. 7.
4 JL Symonds, *A History of British Atomic Tests in Australia*, AGPS, Canberra, 1985, p. 13.
5 Royal Commission into British Nuclear Tests in Australia, *Report* (JR McClelland, president), 2 vols, AGPS, Canberra, 1985, p. 103.
6 William Penney, Epicure. Report on Proposed Site, 12 January 1951. NAA: A6455, BV1.
7 Penney, Epicure report, 13 January 1951.
8 Penney, Epicure report, 13 January 1951.
9 Penney, Epicure report, 13 January 1951.
10 Lorna Arnold and Mark Smith, *Britain, Australia and the Bomb: The Nuclear Tests and Their Aftermath*, 2nd edn, Palgrave Macmillan, New York, 2006, p. 29.

11 Penney, Epicure report, 13 January 1951.
12 Penney, Epicure report, 13 January 1951.
13 Margaret Gowing, 'Monte Bello', in *Independence and Deterrence: Policy Execution*, vol. 2, Palgrave Macmillan, London, 1974, p. 479.
14 Royal Commission *Report*, p. 103.
15 HUREX Signal No.1, via Admiralty to UKSLS Melbourne, 25 June 1951. TNA: DEFE 16/505.
16 Noah Pearce, Statement to the Royal Commission [Royal Commission into British Nuclear Tests in Australia], Statements from UK Witnesses: M–Z]. NAA: A6449, 2.
17 UK Admiralty, Secret message to UKSLS Melbourne, 20 June 1951, Operation Hurricane, copies of correspondence on operational matters between the UK Hurricane Executive and Hurricane Panel (Aust). NAA: A6456 R095/002.
18 Vice Admiral Sir John Collins, UKSLS, Top-secret letter to chief of the Naval Staff, 21 August 1951, Operation Hurricane, copies of correspondence on operational matters between the UK Hurricane Executive and Hurricane Panel (Aust). NAA: A6456 R095/002.
19 Cabinet APEX committee, Minutes of first meeting, No. 10 Downing Street, 26 March 1952. TNA CAB 134/734.
20 Expert Committee on the Review of Data on Atmospheric Fallout Arising from British Nuclear Tests in Australia, *Report* (CB Kerr, Chair) (Kerr Report), 31 May 1984, p. 6.
21 'Churchill in Control of A-bomb Test', Adelaide *News*, 15 May 1952, p. 1.
22 'Churchill in Control of A-bomb Test', May 1952.
23 APEX committee, Minutes of third meeting , No. 10 Downing Street, 24 April 1952.
24 William Penney and VHB Macklen, obituary, William Richard Joseph Cook, 10 April 1905–16 September 1987, <royalsocietypublishing.org/>, accessed 9 January 2024.
25 APEX committee, Minutes of first meeting.
26 Ruane, *Churchill and the Bomb in War and Cold War*, p. 209.
27 Ruane, *Churchill and the Bomb in War and Cold War*, p. 209.
28 Ruane, *Churchill and the Bomb in War and Cold War*, p. 211.
29 Cabinet APEX committee, Minutes of second meeting, Cabinet Office, 15 April 1952. TNA CAB 134/734.
30 APEX committee, Minutes of second meeting.
31 Royal Commission *Report*, p. 105.
32 'Britain joins the atomic powers', *The Age*, 6 October 1952, p. 2.
33 Royal Commission Report, p. 7.
34 Department of Defence, For Press: Atomic Weapon Test (media release), 13 August 1952. NAA: A816, 10/301/129.
35 UK high commissioner, Canberra, Telegram to secretary of state for Commonwealth relations, 30 May 1952. TNA: FCO 1/4.
36 UK high commissioner telegram, 30 May 1952. Note that in the original the word 'aroused' had been crossed out and 'stimulated' placed instead.
37 Huw Dylan, 'Operation TIGRESS: Deception for Counterintelligence and Britain's 1952 Atomic Test', *Journal of Intelligence History*, vol. 14, no. 1, 14 August 2014, p. 3.
38 William Penney, Top-secret note to Leslie Martin, 14 July 1952. TNA: ES 1/498.
39 Dylan, Operation TIGRESS, p. 1.
40 Dylan, Operation TIGRESS, p. 4.
41 Commonwealth Relations Office, Top-secret telegram to high commissioner for the UK, 23 February 1952. NAA: A816, 3/301/534.
42 CRO telegram to UK high commissioner, February 1952.

43 Department of Defence minute for Minister Philip McBride, 8 May 1952. NAA: A816, 3/301/534.
44 EJ Bunting, acting secretary to Cabinet, Secret Cabinet minute, Decision No. 431, Submission No. 272 – Atomic Weapon Test – Security, 27 May 1952. NAA: A816, 3/301/534.
45 Report of the Montebello Working Party on Operation Mosaic, 10 October 1955. Prepared for the full Maralinga Committee. NAA: A6456, R029/207.
46 Alan McNicoll, deputy chief of Naval Staff, Draft advice to UK high commissioner, 28 May 1952. NAA: A816, 3/301/534.
47 Alan McNicoll, Advice to UK high commissioner, 28 May 1952.
48 Ben Cockram, office of the UK high commissioner, Secret letter to Allan McKnight, 13 June 1952. NAA: A816, 3/301/534.
49 Cockram, Letter to McKnight, June 1952.
50 Ben Cockram, office of the UK high commissioner, Secret letter to Allan McKnight, 1 July 1952. NAA: A816, 3/301/534.
51 Cockram, Letter to McKnight, July 1952.
52 AS Brown, secretary, Prime Minister's Department, Secret reply to Ben Cockram, office of the UK High Commissioner, 9 July 1952. NAA: A816, 3/301/534.
53 Sir Kenneth Bailey, solicitor-general, Secret letter to AS Brown, Prime Minister's Department, 18 July 1952. NAA: A816, 3/301/534.
54 'Security at Atom Islands Worries our Navy Chiefs', *Melbourne Herald*, 2 August 1952, p. 3.
55 Hurricane Panel, Remarks on UK High Commission letter of 1 July, undated. NAA: A816, 3/301/534.
56 'Nuclear Radiation', The Atomic Archive, <www.atomicarchive.com/science/effects/radiation>, accessed 7 January 2026.
57 Royal Commission *Report*, p. 50.
58 Symonds, *A History of British Atomic Tests*, p. 36.
59 Royal Commission *Report*, p. 50.
60 William Penney, Memorandum on safety standards for Operation Hurricane, 30 July 1952. NAA: A6455, RC78.
61 Penney, Memo on safety standards for Operation Hurricane.
62 Penney, Memo on safety standards for Operation Hurricane.
63 Royal Commission *Report*, p. 52.

Chapter 3: Naval operations

1 'Atomic Fleet Sails from Fremantle', *West Australian*, 23 April 1952, p. 1.
2 *The Plym Pudding* (author unknown), newsletter, 'Posthumous Report', No. 15, 24 October 1952. In <www.labrats.international/post/the-plym-pudding>, accessed 7 March 2025.
3 Robert Milliken, *No Conceivable Injury: The Story of Britain and Australia's Atomic Cover-Up*, Penguin, Ringwood Vic., 1986, p. 44.
4 Margaret Gowing, 'Monte Bello', in *Independence and Deterrence: Britain and Atomic Energy, 1945–1952*, Palgrave Macmillan, London, 1974, p. 481.
5 Cabinet APEX Committee, Minutes of first meeting, No. 10 Downing Street, 26 March 1952. TNA: CAB 134/734.
6 Milliken, *No Conceivable Injury*, p. 44.
7 *The Plym Pudding* (author unknown), newsletter, 'Opening Number', No. 1, 19 February 1952.
8 Gowing, 'Monte Bello', p. 492.

9 Milliken, *No Conceivable Injury*, p. 45.
10 Gowing, 'Monte Bello', p. 485.
11 David Wilson, *Always First: The RAAF Construction Squadrons 1942–1974*, Air Power Study Centre, Canberra, 1998, p. 103.
12 Wilson, *Always First*, p. 110.
13 Wilson, *Always First*, p. 110.
14 Wilson, *Always First*, p. 111.
15 Wilson, *Always First*, p. 111.
16 'Brisbane Airman Wins Medal at Cocos Island', *Courier-Mail* (Brisbane), 3 November 1952, p. 1.
17 'Brisbane Airman Wins Medal at Cocos Island', November 1952.
18 'After Cocos the R.A.A.F. turns to Townsville', *Townsville Daily Bulletin*, 18 December 1952, p. 6.
19 Gowing, 'Monte Bello', p. 486.
20 Paul Grace, *Operation Hurricane: The Story of Britain's First Atomic Test in Australia and the Legacy that Remains*, Hachette, Sydney, 2023, pp. 121–22.
21 Gowing, 'Monte Bello', p. 486.
22 PW Brooking, Note attached to PF Cooper, Operation Hurricane: Report by Capt. PF Cooper RN (Retd), 4 October 1954. TNA: ES 1/147.
23 Brooking, Note attached to PF Cooper report, October 1954.
24 PF Cooper, Operation Hurricane: Report by Capt. PF Cooper RN (Retd), 4 October 1954. TNA: ES 1/147.
25 Cooper, Operation Hurricane Report, October 1954.
26 PW Brooking, Note to E Evans-Lombe, HUREX, attached to Leonard Tyte, Operation Hurricane: Dr Tyte's Diary, 1952. TNA: ES 1/78.
27 Brooking, Note to E Evans-Lombe, attached to Dr Tyte's Diary, 1952.
28 Leonard Tyte, Operation Hurricane: Dr Tyte's Diary, 1952. TNA: ES 1/78.
29 Tyte, Diary, 1952.
30 Tyte, Diary, 1952.
31 JJ McEnhill, Story of Operation Hurricane, presented 26/7/85 at Sydney, Royal Commission, NAA: A6455, RC603, p. 1.
32 McEnhill, p. 42.
33 McEnhill, p. 42.
34 McEnhill, p. 45.
35 McEnhill, p. 45.
36 McEnhill, p. 52.
37 HUREX, via Admiralty, Top-secret message to Commander Task Force 4 (Torlesse), Order HTO 262, 19 September 1952 . TNA: FCO 1/5.
38 HUREX, Order HTO 262.
39 JL Symonds, *A History of British Atomic Tests in Australia*, AGPS, Canberra, 1985, p. 24.
40 Gowing, 'Monte Bello', pp. 489–90.
41 Cooper, Operation Hurricane Report, October 1954.
42 Cooper, Operation Hurricane Report, October 1954.
43 Symonds, *A History of British Atomic Tests*, p. 75.
44 Symonds, *A History of British Atomic Tests*, p. 75.
45 Symonds, *A History of British Atomic Tests*, p. 76.
46 Symonds, *A History of British Atomic Tests*, p. 76.
47 Philip McBride, minister for defence, Statement, 'Australian Warships and Scientists Present at the Atomic Test', 3 October 1952. TNA: FCO 1/5.

48 Symonds, *A History of British Atomic Tests*, p. 76.
49 Symonds, *A History of British Atomic Tests*, p. 77.
50 Philip McBride, minister for defence, 'Atomic Tests', Top-secret letter to Howard Beale, minister for supply, 8 June 1955. NAA: A6456, R209/005.
51 Norm Milne, Handwritten notes, 17 April 1952. Milne 'Atomic Testing' file, courtesy of Rory Milne.
52 Milne, Handwritten notes, April 1952.
53 Milne, Handwritten notes, April 1952.
54 Milne, Handwritten notes, April 1952.
55 Milne, Handwritten notes, April 1952.
56 'Likely Site of British Bomb Experiment', *West Australian*, 22 April 1952, p. 1.
57 'Possible Site of Atomic Tests', *West Australian*, 22 April 1952, p. 10.
58 Darcy Farrell, 'Douglas J Burton', Doug Burton Tribute, WA TV History, 11 November 2014, <watvhistory.com/2014/11/doug-burton-tribute-part-5-of-6>, accessed 26 March 2025.
59 '2200-Mile Mission to Cover Atom Bomb Story', *The Journalist*, July 1952, p. 2.
60 '2200-Mile Mission to Cover Atom Bomb Story', *The Journalist*, July 1952.
61 Grace, *Operation Hurricane*, p. 97.
62 Grace, *Operation Hurricane*, pp. 97–98.
63 Daily News Special Correspondent (Jack Coulter), 'Pressmen Find Montebellos Like Wartime Beachhead', *Daily News* (Perth), 29 April 1952, p. 2.
64 Daily News Special Correspondent (Jack Coulter), 'Pressmen Find Montebellos Like Wartime Beachhead'.
65 Norm Milne, Memorandum to news editor, 3 June 1952, Milne 'Atomic Testing' file, courtesy of Rory Milne.
66 Milne, Memorandum to news editor, June 1952.
67 Leonard Tyte, Operation Hurricane: Dr Tyte's Diary, 1952. TNA: ES 1/78.
68 UK high commissioner in Australia (acting), Secret telegram to CRO, 28 August 1952. TNA: FCO 1/5.
69 UK high commissioner in Australia (acting), Secret telegram to CRO, 28 August 1952.
70 'The Monte Bello Expedition. A Pictorial Record of the Joint Expedition Undertaken by *The West Australian* and *The Daily News* to Provide a Cover of Britain's First Atomic Explosion – October 1952', BA1716, Sir James Cruthers Collection, JS Battye History Collection, State Library of Western Australia, <slwa.wa.gov.au/pdf/pictorial/BA1716.pdf>, accessed 28 September 2025.
71 'Pressmen's Long Watch on Monte Bello', *The Age* (Melbourne), 4 October 1952, p. 2.
72 'Pressmen's Long Watch', October 1952.
73 'Pressmen's Long Watch', October 1952.
74 'Pressmen's Long Watch', October 1952.
75 Darcy Farrell, 'Douglas J Burton', Doug Burton Tribute, WA TV History.
76 '"Space Suit" Scientists Check on Atom Blast', *Sun News-Pictorial*, Melbourne, 6 October 1952, p. 3; 'The Monte Bello Expedition', October 1952.
77 'Britain Tests Her First Atomic Bomb off W.A. Coast', *West Australian*, 4 October 1952, p. 1.
78 'Newsmen Kept a Long Watch on a Hilltop', *West Australian*, 4 October 1952, p. 11.
79 Mount Potter – Western Australia – Post Office history. NAA: C3629, MOUNT POTTER.
80 Mount Potter – Western Australia – Post Office history.
81 Mount Potter – Western Australia – Post Office history.

82 Mount Potter – Western Australia – Post Office history.
83 Alec Walkling, Statement to the Royal Commission [Royal Commission into British Nuclear Tests in Australia], Statements from UK Witnesses: M–Z. NAA: A6449, 2.
84 Geoffrey Dhenin, Medical Officer's Journal, HMS *Tracker*. NAA: A13070, Smith 167, WE Smith vs Commonwealth of Australia Document 167 (Folder 13).
85 Dhenin, Medical Officer's Journal.
86 Dhenin, Medical Officer's Journal.
87 Dhenin, Medical Officer's Journal.
88 Dhenin, Medical Officer's Journal.
89 Dhenin, Medical Officer's Journal.
90 Dhenin, Medical Officer's Journal.
91 R Mortimer, Medical Officer's Journal, HMS *Tracker*. NAA: A13070, Smith 167, WE Smith vs Commonwealth of Australia Document 167 (Folder 13).
92 JA Hole, Statement to the Royal Commission [Royal Commission into British Nuclear Tests in Australia], Statements from UK Witnesses: A–L]. NAA: A6449, p. 1.
93 JA Hole, Evidence to the Royal Commission, Transcript of Proceedings. NAA: A6448, 10, pp. 5154–55.
94 Hole, Statement to the Royal Commission.
95 Wayne Reynolds, 'Rethinking the Joint Project: Australia's Bid for Nuclear Weapons, 1945–1960', *Historical Journal*, vol. 41, no. 3, 1998, p. 862.
96 Margaret Gowing, 'Monte Bello', p. 477.
97 McEnhill, p. 59.
98 McEnhill, p. 59.
99 McEnhill, p. 59.
100 McEnhill, p. 64.
101 McEnhill, p. 65.
102 McEnhill, p. 65.
103 McEnhill, p. 65.
104 McEnhill, pp. 65–66.

Chapter 4: *Plym* no more

1 *This Little Ship*, Adrian Cooper and Ron Osborn (directors), UK Atomic Energy Authority, 1953, <www.sonicbomb.com/v1.php?vid=content/atomic/ifilms/this_little_ship&h=360&w=480&ttitle=This%20Little%20Ship%201953&d=>, accessed 20 March 2025.
2 JJ McEnhill, Story of Operation Hurricane, Presented 26/7/85 at Sydney, Royal Commission, NAA: A6455, RC603, p. 66.
3 Expert Committee on the Review of Data on Atmospheric Fallout Arising from British Nuclear Tests in Australia, *Report* (CB Kerr, Chair) (Kerr Report), 31 May 1984, p. 6.
4 PF Cooper, Operation Hurricane: Report by Capt. PF Cooper RN (Retd), 4 October 1954. TNA: ES 1/147.
5 *This Little Ship*, Cooper and Osborn, 1953.
6 *This Little Ship*, Cooper and Osborn, 1953.
7 *Operation Hurricane*, film directed by Ronald Stark, produced by Stuart Legg, UK Ministry of Supply, 1952, <www.iwm.org.uk/collections/item/object/1060022141>, accessed 25 March 2025.
8 Helen Hughes, 'Film and the British Atomic Project', *Historical Journal of Film, Radio and Television*, vol. 42, no. 2, 2022, p. 227.
9 WE Duckworth, 'Ieuan Maddock: 29 March 1917–29 December 1988', Biographical

Memoirs of Fellows of the Royal Society, 1 November 1991, <royalsocietypublishing.org/doi/abs/10.1098/rsbm.1991.0016>, accessed 13 January 2025.

10 Duckworth, 'Ieuan Maddock'.

11 Duckworth, 'Ieuan Maddock'.

12 Royal Commission into British Nuclear Tests in Australia, *Report* (JR McClelland, president), 2 vols, AGPS, Canberra, 1985, p. 123.

13 Michael Stephens, Statement to the Royal Commission [Royal Commission into British Nuclear Tests in Australia], Statements from UK Witnesses: M–Z. NAA: A6449, p. 2.

14 Margaret Gowing, 'Monte Bello', in *Independence and Deterrence: Britain and Atomic Energy, 1945–1952*, Palgrave Macmillan, London, 1974, p. 494.

15 Sean-Paul Stephens and Ian Wally, 'Poignant Tribute to Dorrie: Memories of Pilbara Massacre, Nuclear Bombs and the Last Marthudunera Speaker', *Pilbara News*, 9 September 2025.

16 Mardie Station, <inherit.dplh.wa.gov.au/public/inventory/printsinglerecord/2498a13e-726f-41f6–81cc-a06b0a916681>, accessed 23 June 2025.

17 Dorrie Walley, oral history, Burrup Stories Project, March 2010. Creator: Kate Gregory. Australian Institute of Aboriginal and Torres Strait Islander Studies, GREGORY_KOO1.

18 Frederick Lindemann, Note to Churchill, 17 September 1952, Lindemann/Cherwell papers J97, Nuffield College Library, Oxford.

19 Lindemann, Note to Churchill, 17 September 1952.

20 Lindemann, Note to Churchill, 17 September 1952.

21 'Dr. Penney Knighted', *Telegraph* (Brisbane), 24 October 1952, p. 1.

22 'Sir W Penney to Broadcast on Atomic Blast', *Canberra Times*, 4 November 1952, p. 1.

23 William Penney, Final Draft of Dr Penney's Broadcast, October 1952. TNA: FCO 1/5.

24 Robert Menzies, prime minister, Statement, 'Atomic Weapon Test', 3 October 1952. TNA: FCO 1/5.

25 William Penney, Operation Hurricane – the dose-rate contours of the residual radioactive contamination, top-secret section of Director's Report, 27 August 1954. TNA: ES5/3.

26 Radiation Sickness, Medline, <medlineplus.gov/ency/article/000026.htm>, accessed 10 February 2025.

27 Penney, Operation Hurricane – dose-rate contours of the residual radioactive contamination, August 1954.

28 Penney, Operation Hurricane – dose-rate contours of the residual radioactive contamination, August 1954.

29 Royal Commission *Report*, p. 117.

30 Royal Commission *Report*, p. 117.

31 Royal Commission *Report*, p. 117.

32 DE Barnes, Fall Out of Contamination for the Monte-Bello Explosion: Criteria for Safe Firing, AWRE report T1/53, May 1953. Some preliminary results from the Monte Bello tests, relevant to defence, including civil defence. NAA: A6455, RC604.

33 Barnes, Fall Out of Contamination for the Monte-Bello Explosion, May 1953.

34 Penney, Broadcast, October 1952.

35 DG Stevenson, Statement to the Royal Commission [Royal Commission into British Nuclear Tests in Australia], Statements from UK Witnesses: M–Z]. NAA: A6449, 2.

36 Stevenson, Statement to the Royal Commission.

37 Reginald Garbett, Submission tendered to the Royal Commission. NAA: A6455, RC470.

38 Office of the UK High Commissioner, Secret note to Allen Brown, Prime Minister's Department Canberra, 22 October 1952. TNA: FCO 1/5.
39 Winston Churchill, prime minister, House of Commons Hansard, vol. 505, Thursday 23 October 1952.
40 Churchill, House of Commons Hansard, October 1952.
41 Churchill, House of Commons Hansard, October 1952.
42 Churchill, House of Commons Hansard, October 1952.
43 Sydney Silverman, Question to the prime minister, House of Commons Hansard, vol. 505, Thursday 23 October 1952.
44 Penney, Broadcast, October 1952.
45 Penney, Broadcast, October 1952.
46 Penney, Broadcast, October 1952.
47 Penney, Broadcast, October 1952.
48 Penney, Broadcast, October 1952.
49 EK Peck, Evidence to the Royal Commission, Transcript of Proceedings. NAA: A6450, 4, p. 880.
50 Henry Carter, Statement to the Royal Commission [Royal Commission into British Nuclear Tests in Australia], Statements from UK Witnesses: A–L]. NAA: A6449, 1, p. 1.
51 Henry Carter, Evidence to the Royal Commission, Transcript of Proceedings. NAA: A6448, 10, p. 5145.
52 Carter, Statement to the Royal Commission, p. 2.
53 Carter, Statement to the Royal Commission, p. 2.
54 Carter, Statement to the Royal Commission, p. 3.
55 Carter, Evidence to the Royal Commission, p. 5138.
56 Carter, Evidence to the Royal Commission, pp. 5141–42.
57 Andrew Thomas, Statement and Weekly Orders issue number 1053, 1/11/1954 and other orders and correspondence, Royal Commission into British Nuclear Tests in Australia, Submission to the Royal Commission, NAA: A6455, RC553, pp. 3–4.
58 Thomas, Statement to the Royal Commission, p. 4.
59 Thomas, Statement to the Royal Commission, p. 5.
60 Thomas, Statement to the Royal Commission, p. 6.
61 Cooper, Operation Hurricane Report, October 1954.
62 Penney, Broadcast, October 1952.
63 Thomas, Statement to the Royal Commission, p. 8.
64 Thomas, Statement to the Royal Commission, p. 9.
65 Thomas, Statement to the Royal Commission, p. 10.
66 Thomas, Statement to the Royal Commission, p. 10.
67 Alec Walkling, Operation Hurricane: Report of the Radiological Hazard Division (draft), Fort Halstead, 15 May 1953. NAA: A6455, RC250.
68 Graham Mabbutt, Statement to the Royal Commission [Royal Commission into British Nuclear Tests in Australia], Statements from UK Witnesses: M–Z. NAA: A6449, 2, pp. 4–5.
69 Mabbutt, Statement to the Royal Commission, p. 6.
70 Mabbutt, Statement to the Royal Commission, p. 6.
71 Alec Walkling, Evidence to the Royal Commission, Transcript of Proceedings. NAA: A6448, 9, p. 4530.
72 Walkling, Evidence to the Royal Commission, p. 4530.
73 'Nuclear Test Inquiry: Radioactive Waste "Harmless"', *Canberra Times*, 23 January 1985, p. 12.

74 AWRE report T1/53, May 1953. Some preliminary results from the Monte Bello tests, relevant to defence, including civil defence. NAA: A6455 RC604, T1/53, p. 5.
75 AWRE report T1/53, p. 5.
76 AWRE report T1/53, p. 6.
77 AWRE report T1/53, p. 8.
78 'Monte Bello Expedition Reaches Onslow', *West Australian*, 9 November 1953, p. 8.
79 Frank Furlong, Western Australian Birth, Deaths and Marriages, <wabmd.toolforge.org/death/1953/101168?utm_source=chatgpt.com - gsc.tab=0>, accessed 24 September 2025.
80 'Monte Bello Expedition Reaches Onslow', *West Australian*, 9 November 1953, p. 8.
81 William Penney, Evidence to the Royal Commission, Transcript of Proceedings. NAA: A6449, 2, p. 4466.
82 Captain FB Lloyd, Secret note to Vice Admiral EM Evans-Lombe, deputy chief of Naval Staff, Admiralty, 31 October 1952. NAA: A6455, AG10.
83 Captain FB Lloyd, Secret note to Neil Pritchard, Commonwealth Relations Office, 2 January 1953. NAA: A6455, AG10.
84 Arthur Andrews, Royal Australian Navy service record. NAA: A6769.
85 Arthur Andrews, Evidence to the Royal Commission, Transcript of Proceedings. NAA: A6448, 4, p. 1319.
86 Andrews, Evidence to the Royal Commission, p. 1323.
87 EK Peck, Statement to the Royal Commission [Royal Commission into British Nuclear Tests in Australia] Statements from Australian Witnesses: P–S. NAA: A6450, 4.
88 EK Peck, Evidence to the Royal Commission, Transcript of Proceedings. NAA: A6450, 4, p. 851.
89 Tas Browning, 'HMAS Hawkesbury', *Sabretache*, vol. LI, no. 1, March 2010, p. 47.
90 Robert Scrivenor, lieutenant, HMAS *Hawkesbury*, Confidential message to the naval officer-in-charge West Australian Area, 30 November 1952. NAA: A6456, R015/011.
91 Scrivenor, Message to naval officer-in-charge, November 1952.
92 Naval officer-in-charge West Australia Area, Department of Navy, Secret cable to HMAS *Hawkesbury*, December 1952. NAA: A6456, R015/011.
93 Naval officer-in-charge West Australian Area, Cable to *Hawkesbury*, December 1952.
94 Naval officer-in-charge West Australian Area, Confidential note to the secretary, Naval Board, Melbourne 4 December 1952. NAA: A6456, R015/011.
95 Naval officer-in-charge West Australian Area, Note to Naval Board, December 1952.
96 Department of the Navy, Secret minute paper, 18 December 1952. NAA: A6456, R015/011.
97 Department of the Navy, Secret minute paper, December 1952.
98 Naval officer-in-charge West Australian Area, 'Valuable scientific stores recovered from the Monte Bello Area', Letter to the secretary, Naval Board, Naval Office Melbourne, 20 March 1953. NAA: A6456, R015/011.
99 Tas Browning, 'HMAS Hawkesbury', *Sabretache*, vol. LI, no. 1, March 2010, p. 48.
100 AS Brown, secretary, Prime Minister's Department, Secret letter to the Admiralty, 9 November 1954. TNA: AVIA 65/1151.
101 Brown, Letter to the Admiralty, 9 November 1954.
102 Brown, Letter to the Admiralty, 9 November 1954.
103 Sir Frederick Morgan, Secret letter to William Penney, 31 December 1952. TNA: ES 38/ZDRS6.
104 TC Bray, 'And Monte Bello is Still "Hot"', Brisbane *Courier-Mail*, 6 August 1953, p. 1.
105 Bray, *Courier-Mail*, August 1953.

106 William Penney, interview, 'Atom Man's Hush-Hush Return: Dr. Penney Returns From Monte Bello', British Pathé, 20 October 1952, <www.youtube.com/watch?v=nbRqFhLCic4>, accessed 20 March 2025.
107 *Operation Hurricane*, film.

Chapter 5: At the crossroads of fusion

1 Churchill quoted by Lord Hennessy of Nympsfield, 'Cabinets and the Bomb', The Inaugural Michael Quinlan Lecture, Queen's Robing Room, House of Lords, Wednesday 2 February 2011.
2 HV Evatt, leader of the opposition, and Howard Beale, minister for supply, Questions, House of Representatives Hansard, 15 February 1956.
3 Stephen Holmes, British high commissioner to Australia, Top-secret telegram to Alec Douglas-Home, secretary of state for Commonwealth relations, 16 March 1955. TNA: FCO 1/8.
4 Expert Committee on the Review of Data on Atmospheric Fallout Arising from British Nuclear Tests in Australia, *Report* (CB Kerr, Chair) (Kerr Report), 31 May 1984, p. 33, <inis.iaea.org/collection/NCLCollectionStore/_Public/17/019/17019465.pdf>, accessed 26 July 2021.
5 Richard Gott, 'The Evolution of the Independent British Deterrent', *International Affairs* (Royal Institute of International Affairs), vol. 39, no. 2, April 1963, p. 238.
6 John Baylis, 'The Development of Britain's Thermonuclear Capability 1954–61: Myth or Reality?', *Contemporary Record*, vol. 8, no. 1, Summer 1994, p. 164.
7 Jim Gledhill, 'The Bow and Arrow Versus the Atom Bomb: Air Defence in Scotland 1945–1955', *British Journal for Military History*, vol. 10, issue 1, March 2024, p. 157.
8 Norman Dombey and Eric Grove, 'Britain's Thermonuclear Bluff', *London Review of Books*, vol. 14, no. 20, 22 October 1992, <www.lrb.co.uk/the-paper/v14/n20/norman-dombey/britain-s-thermonuclear-bluff>, accessed 27 March 2025.
9 EC Williams, Top-secret note for Air Defence Committee Working Party, 'Some Speculations about the Effects of Developments in Atomic Weapons Upon Future Defence Policy, with a Hypothesis about the Consequences to the Provision of Air Defence for the United Kingdom', 8 May 1954. TNA: DEFE 7/2208.
10 Williams, 'Some Speculations about the Effects of Developments in Atomic Weapons', May 1954.
11 Williams, 'Some Speculations about the Effects of Developments in Atomic Weapons', May 1954.
12 Robert Standish Norris, 'Questions on the British H-Bomb', Natural Resources Defense Council, 22 June 1992, p. 1.
13 Albert Blackburn, Question to the secretary of state for foreign affairs, House of Commons Hansard, vol. 472, Wednesday 22 March 1950.
14 Ernest Bevin, secretary of state for foreign affairs, Answer to question from Mr Blackburn, House of Commons Hansard, vol. 472, Wednesday 22 March 1950.
15 I Montgomery, Ministry of Defence, Letter to Frank How, ministry of supply, 8 March 1950. TNA: DEFE 7/2208.
16 Jan Melissen, 'Prelude to Interdependence: The Anglo-American Relationship and the Limits of Great Britain's Nuclear Policy, 1952–1957', *Contemporary Security Policy*, vol. 11, no. 3, 1990, p. 215.
17 Melissen, 'Prelude to Interdependence', p. 215.
18 Lord Hennessy of Nympsfield, 'Cabinets and the Bomb', February 2011.
19 John R Walker, *British Nuclear Weapons and the Test Ban 1954–1973*, Ashgate Publishing, Farnham, 2010, p. 10.

20 Robert Standish Norris, 'Questions on the British H-Bomb', Natural Resources Defense Council, 22 June 1992, p. 2.
21 Dombey and Grove, 'Britain's Thermonuclear Bluff'.
22 Norman Dombey, 'John Clive Ward, 1 August 1924–6 May 2000', obituary, Biographical Memoirs of the Fellows of the Royal Society, 17 February 2021, <royalsocietypublishing.org/doi/10.1098/rsbm.2020.0023>, accessed 27 March 2025.
23 PH Dean, Foreign Office, Top-secret letter to Sir Frederick Brundrett, Ministry of Defence, 8 July 1945. TNA: DEFE 7/897.
24 Dean, Letter, 8 July 1945.
25 Dean, Letter, 8 July 1945.
26 Sir Frederick Brundrett, Ministry of Defence, Top-secret letter to PH Dean, Foreign Office, 13 July 1954. TNA: DEFE 7/897.
27 Lord Salisbury, Top-secret letter to Field Marshal Earl Alexander of Tunis, 20 August 1954. TNA: DEFE 7/897.
28 Roger Makins, Foreign Office, Top-secret record of conversation with Admiral Lewis Strauss, 10 March 1955. TNA: DEFE 7/897.
29 Makins, Record of conversation with Strauss, March 1955.
30 Makins, Record of conversation with Strauss, March 1955.
31 Ministry of Defence briefing note, Nuclear Tests, 18 July 1955. TNA: DEFE 7/897.
32 Ministry of Defence briefing note, Nuclear Tests, July 1955.
33 Ministry of Defence briefing note, Nuclear Tests, July 1955.
34 Ministry of Defence briefing note, Nuclear Tests, July 1955.
35 Katherine Pyne, 'Art or Article? The Need for and Nature of the British Hydrogen Bomb, 1954–58', *Contemporary British History*, vol. 9, no. 3, p. 568.
36 Davey, T Division, Home Office, Civil Defence Department, Training: Memorandum on effects of the hydrogen bomb, 20 August 1954. TNA: HO 338/23.
37 Pyne, 'Art or Article?', p. 568.
38 Secretary of state for Commonwealth relations, Top-secret telegram to high commissioner for the UK, Australia, 2 April 1955. TNA: FCO/1/8.
39 Pyne, 'Art or Article?', p. 573.
40 Pyne, 'Art or Article?', p. 569.
41 Pyne, 'Art or Article?', p. 572.
42 Pyne, 'Art or Article?', p. 563.
43 Margaret Gowing, with Lorna Arnold, *Independence and Deterrence: Britain and Atomic Energy, 1945–1952*, United Kingdom Atomic Energy Authority, St Martin's Press, New York, 1974, p. 450.
44 Kevin Ruane, *Churchill and the Bomb in War and Cold War*, Bloomsbury Academic, London, 2016, p. 265.
45 Pyne, 'Art or Article?', p. 571.
46 Lorna Arnold and Mark Smith, *Britain, Australia and the Bomb: The Nuclear Tests and Their Aftermath*, 2nd edn, Palgrave Macmillan, New York, 2006, p. 76.
47 Ruane, *Churchill and the Bomb*, p. 415.
48 Winston Churchill, Top-secret cabinet document, 14 December 1954: 'Notes on Tube Alloys, 1954'. DSND 14/4 Churchill Archives Centre, University of Cambridge. .
49 Walker, *British Nuclear Weapons and the Test Ban 1954–1973*, p. 5.
50 Henry Pelling, 'The H-Bomb and Eden's Diplomacy', in *Churchill's Peacetime Ministry, 1951–55*, Palgrave Macmillan, London, 1997, p. 120.
51 Winston Churchill, prime minister, Statement on Defence, House of Commons Hansard, vol. 537, Tuesday 1 March 1955.
52 Churchill, Statement on Defence, House of Commons, March 1955.

53 David Reynolds, 'A "Special Relationship"? America, Britain and the International Order Since the Second World War', *International Affairs* (Royal Institute of International Affairs 1944–), vol. 62, no. 1, Winter 1985–1986, p. 9.
54 Anthony Eden, prime minister, Debate, Atom and Hydrogen Bomb Experiments, House of Commons Hansard, vol. 547, Tuesday 6 December 1955.
55 Anthony Eden, Hansard, December 1955.
56 The Rand Corporation, 'Worldwide Effects of Atomic Weapons: Project Sunshine', ASTIA Document Number AD 112389, 6 August 1953, amended 1956, Santa Monica, California, p. 3.
57 Rand Corporation, 'Worldwide Effects of Atomic Weapons: Project Sunshine', p. 3.
58 Rand Corporation, 'Worldwide Effects of Atomic Weapons: Project Sunshine', p. 3.
59 Walker, *British Nuclear Weapons and the Test Ban 1954–1973*, p. 13.
60 William Penney and VHB Macklen, obituary, William Richard Joseph Cook, 10 April 1905–16 September 1987, <royalsocietypublishing.org/rsbm/article/doi/10.1098/rsbm.1988.0003/88615/William-Richard-Joseph-Cook-10-April-1905-16>, accessed 9 January 2024.
61 Anthony Eden, UK prime minister, personal note to Robert Menzies, Australian prime minister, 16 May 1955. NAA: A6456, R084/001.
62 Eden, note to Menzies, May 1955.
63 Eden, note to Menzies, May 1955.
64 Robert Menzies, Australian prime minister, Letter to Anthony Eden, UK prime minister, 20 June 1955. NAA: A6456, R084/001.
65 Menzies note to Eden, June 1955.
66 Anthony Eden, UK prime minister, Note to Robert Menzies, Australian prime minister, 8 July1955. NAA: A6456, R084/001.
67 Eden note to Menzies, July 1955.
68 Eden note to Menzies, July 1955.
69 Anthony Eden, prime minister, Statement to the House of Commons, Hansard, vol. 552, 16 May 1956.
70 Alec Douglas-Home, secretary of state for Commonwealth affairs, Personal telegram to Anthony Eden, prime minister, 4 December 1956. TNA: CAB 21/3148.
71 Douglas-Home, Personal telegram to Eden, December 1956.
72 Alec Douglas-Home, secretary of state for Commonwealth affairs, Draft telegram to the prime minister of Australia and the prime minister of New Zealand, 12 December 1956. TNA: CAB 21/3148.
73 Emrys Hughes, House of Commons Hansard, vol. 554, Monday 11 June 1956, p. 21.
74 Reginald Maudling, minister of supply, House of Commons Hansard, vol. 554, Monday 11 June 1956, p. 21.
75 Stephen Swingler, Debates, House of Commons Hansard, vol. 554, Tuesday 19 June 1956.
76 Anthony Eden, prime minister, Debates, House of Commons Hansard, vol. 554, Tuesday 19 June 1956.

Chapter 6: Australians at Monte Bello

1 UKSLS, letter to Ben Cockram, 24 September 1952. Telegram (HUREX 115) to UKSLS: Martin. TNA: FCO 1/5/11.
2 'Safe in London ...', Sydney *Tribune*, 27 June 1956, p. 1.
3 'Atomic History', *West Australian*, 4 October 1952, p. 2.
4 'Atomic History', p. 2.
5 Royal Commission into British Nuclear Tests in Australia, *Report: Conclusions and*

Recommendations (JR McClelland, president, J Fitch and WJA Jonas, commissioners), AGPS, Canberra, 1985, p. 106.

6 WM Bourke, Questions, House of Representatives Hansard, 29 August 1952.

7 Philip McBride, Questions, House of Representatives Hansard, 29 August 1952.

8 William Penney, Secret note to Frederick Lindemann (Lord Cherwell), 27 February 1952. TNA: ES 1/498.

9 Penney, Note to Lindemann, 27 February 1952.

10 EL Sykes, Commonwealth Relations Office, Secret letter to Frank How, Ministry of Supply. TNA: ES 1/498.

11 Defence Committee, Minutes of meeting, 28 August 1952. NAA: A6456, R096/006.

12 George Davey, British High Commission Canberra, Secret letter to Allan McKnight, deputy secretary, Prime Minister's Department, 27 August 1952. NAA: A6456, R096/006.

13 Defence Committee, Minutes of meeting, 4 September 1952. NAA: A6456, R096/006.

14 William Penney, Letter to Frederick Lindemann (Lord Cherwell), 15 August 1952. NAA: A6455, RC468.

15 EJS Clarke, Office of the Paymaster-General, Letter to the CRO, 17 September 1952. TNA: FCO 1/5.

16 FC How, Ministry of Supply, Top-secret letter to RM Saner, Foreign Office, 19 August 1952. TNA: FCO 1/4.

17 How, Letter to Saner, August 1952.

18 Neil Pritchard, Commonwealth Relations Office, Top-secret letter to RM Saner, 19 August 1952. TNA: FCO 1/4.

19 William Penney, Top-secret message to Leslie Martin, 22 August 1952. TNA: FCO 1/4.

20 UKSLS, Letter to Cockram, September 1952.

21 UKSLS, Letter to Cockram, September 1952.

22 Defence Committee minutes, 4 September 1952.

23 Tim Sherratt, 'A Political Inconvenience: Australian Scientists at the British Atomic Weapons Tests, 1952–53', *Historical Records of Australian Science*, vol. 6, no. 2, December 1985, p. 146.

24 Sherratt, 'A Political Inconvenience', p. 147.

25 UKSLS, Letter to Cockram, September 1952.

26 UKSLS, Letter to Cockram, September 1952.

27 UKSLS, Letter to Cockram, September 1952.

28 Sir John Cockcroft, Handwritten note to William Penney, 6 September 1952. TNA: ES 1/498.

29 Ernest Titterton, Evidence to the Royal Commission, Transcript of Proceedings. NAA: A6449, 14, p. 7617.

30 Titterton, Evidence to the Royal Commission, p. 7617.

31 JO Newton and John Jenkin, 'Sir Ernest William Titterton 1916–1990', *Australian Dictionary of Biography*, <adb.anu.edu.au/biography/titterton-sir-ernest-william-973>, accessed 25 September 2025.

32 Titterton, Evidence to the Royal Commission, p. 7618.

33 Titterton, Evidence to the Royal Commission, p. 7619.

34 Titterton, Evidence to the Royal Commission, p. 7619.

35 Titterton, Evidence to the Royal Commission, p. 7620.

36 AS Brown, secretary, Prime Minister's Department, Top-secret memo to Frederick Shedden, secretary, Department of Defence, 23 September 1952. NAA: A6456, R095/004.

37 Sherratt, 'A Political Inconvenience', p. 143.
38 Brian Toohey, *Secret: The Making of Australia's Security State*, Melbourne University Press, Melbourne, 2019, p. 61.
39 William Penney, Evidence to the Royal Commission, Transcript of Proceedings. NAA: A6449, 2, p. 4327.
40 Penney, Evidence to the Royal Commission, p. 4327.
41 Captain FB Lloyd, secret memorandum to William Penney, 7 January 1955. TNA: AVIA 65/1151.
42 Gurth Kimber, UK High Commission in Australia, Secret memo to Frank Lloyd, 28 July 1955. TNA: FCO 1/8.
43 Kimber, Memo to Frank Lloyd.
44 William Cook, Atomic Weapons Research Establishment, Top-secret letter to Sir Richard Powell, Ministry of Defence, 25 April 1955. TNA: ES 38/ZFGWL.
45 Cook, Top-secret letter to Sir Richard Powell, 25 April 1955.
46 Atomic Weapons Tests Safety Committee, Minutes of Fourth Meeting, University of Melbourne, 10 January 1956. NAA: A6456, R120/137 PART 1.
47 AWTSC, Minutes of Fourth Meeting, January 1956.
48 AWTSC, Minutes of Fourth Meeting, January 1956.
49 AWTSC, Minutes of Fourth Meeting, January 1956.
50 HA Wills, AWTSC secretary, Memorandum to Leonard Dwyer, Commonwealth Bureau of Meteorology, 2 May 1956. NAA: A6456, R120/137 PART 1.
51 Wills, Memorandum to Dwyer, May 1956.
52 Atomic Weapons Tests Safety Committee, Minutes of Eighth Meeting, University of Melbourne, 23 July 1956. NAA: A6456, R120/137 PART 1.
53 Ernest Titterton, 'Why Australia Is Preferred as an Atom Testing Site', *Sydney Morning Herald*, 16 May 1956, p. 2.
54 Lorna Arnold and Mark Smith, *Britain, Australia and the Bomb: The Nuclear Tests and Their Aftermath*, 2nd edn, Palgrave Macmillan, New York, 2006, p. 123.
55 Martell, quoted in Arnold and Smith, *Britain, Australia and the Bomb*, p. 123.
56 Martell, quoted in Arnold and Smith, *Britain, Australia and the Bomb*, p. 124.
57 Martell, quoted in Arnold and Smith, *Britain, Australia and the Bomb*, p. 124.
58 Royal Commission into British Nuclear Tests in Australia, *Report* (JR McClelland, president), 2 vols, AGPS, Canberra, 1985, pp. 482–83.
59 Royal Commission *Report*, p. 483.
60 JL Symonds, *A History of British Atomic Tests in Australia*, AGPS, Canberra, 1985, p. 336.
61 'Protests Flood in as H-hour Nears: Stop Bomb Tests They Say', *Tribune* (Sydney), 25 April 1956, p. 3.
62 'Protests Flood in as H-hour Nears', *Tribune*, 25 April 1956, p. 3.
63 'Labor Will Stop Weapons Test', *Canberra Times*, 25 June 1956.
64 'A-cloud Adrift – But no Danger', *Argus* (Melbourne), 21 June 1956, p. 1.
65 'A-cloud Adrift', June 1956.
66 Royal Commission *Report*, p. 487.
67 Royal Commission *Report*, p. 487.
68 Royal Commission *Report*, p. 488.
69 'Threat to 3 Towns: Atom Cloud Drift', *Sun* (Sydney), 21 June 1956, p. 1.
70 'Threat to 3 Towns', 21 June 1956, p. 1.
71 UK high commissioner in Australia, Confidential telegram to Commonwealth Relations Office, 21 June 1956. TNA: ES 1/489.
72 UK high commissioner, Telegram to CRO, 21 June 1956.

73 Royal Commission *Report*, p. 487.
74 Royal Commission *Report*, p. 487.
75 HV Evatt, leader of the opposition, Questions, House of Representatives Hansard, 21 June 1956, p. 3515.
76 Evatt, Hansard, 21 June 1956, p. 3515.
77 AW Fadden, acting prime minister, Questions, House of Representatives Hansard, 21 June 1956, p. 3515.
78 Fadden, Hansard, 21 June 1956, p. 3515.
79 Evatt, Hansard, 21 June 1956, p. 3515.
80 Fadden, Hansard, 21 June 1956, p. 3515.
81 Department of Supply, Press Reaction to Atomic Trials, briefing document, 1956 (exact date unknown). NAA: A6456, R047/011.
82 Department of Supply, Press Reaction to Atomic Trials.
83 Department of Supply, Press Reaction to Atomic Trials.
84 Howard Beale, minister for supply, 'Why We Hold A-tests in Australia', 6 August 1956. NAA: A6456, R087/088.
85 GW Tory, British deputy high commissioner Canberra, Minutes of meeting between Adams, Martell, Tory and Rouse, 28 July 1956. NAA: A6455, RC267.
86 Tory, Minutes of meeting, July 1956.
87 Tory, Minutes of meeting, July 1956.
88 Tory, Minutes of meeting, July 1956.
89 Tory, Minutes of meeting, July 1956.
90 Tory, Minutes of meeting, July 1956
91 Report of the Montebello Working Party on Operation Mosaic, 10 October 1955. Prepared for the full Maralinga Committee. NAA: A6456, R029/207.
92 DR Serpell, Draft letter to Jenkyns, Treasury representative in Australia, 13 July 1955. TNA, FCO 1/8.
93 Serpell draft letter to Jenkyns.
94 Serpell draft letter to Jenkyns.
95 Serpell draft letter to Jenkyns.
96 William Penney, Transcript of proceedings, Royal Commission into British Nuclear Tests in Australia, NAA: A6448, 12, p. 7036.
97 Penney, Royal Commission Transcript of proceedings, p. 7037.

Chapter 7: Piecing together Mosaic

1 Sir Frederick Brundrett, Working Party on the Operational Use of Atomic Weapons, Memorandum by the chairman, presented to the meeting of Chiefs of Staff, September 1954. TNA: DEFE 7/2208.
2 Admiralty, Top-secret note about Operation Mosaic, Office of D.C.N.S, 21 June 1955, presented at Executive for Mosaic ad-hoc meeting on 6 July 1955 at the Ministry of Supply. TNA: FCO 1/8.
3 Admiralty, Top-secret note about Operation Mosaic, June 1955.
4 Admiralty, Top-secret note about Operation Mosaic, June 1955.
5 Hugh Martell, Mosaic – Monte Bello Atomic Tests, 1956. Report by Operational Commander. NAA: A6455, RC233, p. 14.
6 Mosaic Executive (MOSEX), Minutes of first meeting, Giles Court London, 6 July 1955. TNA: ADM 296/5.
7 Admiralty, Top-secret note about Operation Mosaic, June 1955.
8 Admiralty, Top-secret note about Operation Mosaic, June 1955.
9 Martell, Mosaic – Monte Bello Atomic Tests, 1956, p. 1.

10 Denis Warner, 'British Atomic Test Explosion Successful: Step Towards Hydrogen Bomb: "Trigger" Experiment', *Daily Telegraph*, 17 May 1956.
11 William Penney, Transcript of proceedings, Royal Commission into British Nuclear Tests in Australia, NAA: A6448, 12, p. 7059.
12 Penney, Royal Commission transcript of proceedings, p. 7059.
13 Martell, Mosaic – Monte Bello Atomic Tests, 1956, p. 1.
14 Martell, Mosaic – Monte Bello Atomic Tests, 1956, p. 1.
15 Martell, Mosaic – Monte Bello Atomic Tests, 1956, p. 1.
16 Report of the Montebello Working Party on Operation Mosaic, 10 October 1955. Prepared for the full Maralinga Committee. NAA: A6456, R029/207, Monte Bello Working Party.
17 Montebello Working Party Report, October 1955.
18 Philip McBride, minister for defence, Top-secret letter to Josiah Francis, minister for the navy, 20 July 1955. NAA: A6456, R031/007.
19 McBride, Letter to Francis, July 1955.
20 MOSEX, meeting minutes, July 1955.
21 MOSEX, meeting minutes, July 1955.
22 MOSEX, meeting minutes, July 1955.
23 Montebello Working Party Report, October 1955.
24 Montebello Working Party Report, October 1955.
25 Martell, Mosaic – Monte Bello Atomic Tests, 1956, p. 16.
26 Montebello Working Party Report, October 1955.
27 Maralinga Sub-Committee, Top-secret report on discussions with United Kingdom officials concerning the support desired from Australia in connection with Operation Mosaic, 1955. NAA: A6456, R031/007.
28 Maralinga Sub-Committee, Report on discussions with United Kingdom officials, 1955.
29 Montebello Working Party Report, October 1955.
30 David McKnight, *Australia's Spies and their Secrets*, Allen & Unwin, Sydney, 1994, p. ix.
31 Minutes of the first meeting of the Monte Bello Working Party held at Navy Office, St Kilda Road, Melbourne, on Wednesday 14 September 1955. NAA: A6456, R029/207.
32 Monte Bello Working Party, Minutes of first meeting, September 1955.
33 Frank O'Connor, Letter to Brigadier CCF Spry, director-general of security, 28 November 1956. NAA: A6456, R087/059.
34 Mosaic Mission to Australia – Summary of Actions, presented to MOSEX Committee, London, 17 August 1955. TNA: ADM 296/12.
35 Mosaic Mission to Australia – Summary of Actions, August 1955.
36 Mosaic Mission to Australia – Summary of Actions, August 1955.
37 William Penney, Transcript of evidence to the Royal Commission into British Nuclear Tests in Australia, NAA A6448, 9, p.4305.
38 AS Brown, secretary, Prime Minister's Department, Top-secret memo to the secretary of the Department of Supply, 13 September 1955. NAA: A6456, R029/207, Monte Bello Working Party – Minutes of meetings, reports, etc.
39 Minutes of the first meeting of the Monte Bello Working Party, 14 September 1955.
40 Minutes of the first meeting of the Monte Bello Working Party, 14 September 1955.
41 Martell, Mosaic – Monte Bello Atomic Tests, 1956, p. 19.
42 Frank O'Connor, secretary, Department of Supply, Top-secret briefing note for Howard Beale, minister for supply, 26 May 1955. NAA: A6456, R209/005.
43 Royal Commission into British Nuclear Tests in Australia, *Report: Conclusions and Recommendations* (JR McClelland, president, J Fitch and WJA Jonas, commissioners), AGPS, Canberra, 1985, p. 482.

44 Royal Commission *Report*, p. 482.
45 EP Hicks, AWRE Aldermaston, Minute paper to Admiral Brooking, 14 October 1954. TNA: ES 1/792.
46 Hicks, Minute paper to Brooking, October 1954.
47 CA Adams, Top-secret letter to Dr DH Black, UKMOSS Melbourne, 29 September 1955. TNA: ADM 296/1.
48 Albert Matthewman, 'Operation Mosaic: Theoretical Predictions', AWRE Report No. T24/57, 15 April 1957. NAA: A6455, RC270.
49 Matthewman, 'Operation Mosaic: Theoretical Predictions', April 1957.
50 Adams, Top-secret letter to Dr DH Black, September 1955.
51 JL Symonds, *A History of British Atomic Tests in Australia*, AGPS, Canberra, 1985, p. 350.
52 Minutes of the third meeting of the Monte Bello Working Party held at 339 Swanston Street, Melbourne, on Friday 30 September 1955. NAA: A6456, R029/207, Monte Bello Working Party – Minutes of meetings, reports, etc.
53 Montebello Working Party Report, October 1955.
54 Montebello Working Party Report, October 1955.
55 Montebello Working Party Report, October 1955.
56 Minutes of the seventh meeting of the Monte Bello Working Party held at 339 Swanston Street, Melbourne, on Friday 18 January 1956. NAA: A6456, R029/207.
57 Minutes of the ninth meeting of the Monte Bello Working Party held at 339 Swanston Street, Melbourne, on Friday 16 April 1956. NAA: A6456, R029/207.
58 Minutes of the ninth meeting of the Monte Bello Working Party, April 1956.
59 Commonwealth Relations Office, Top-secret telegram to UK High Commission in Australia, 13 June 1956. TNA, ES 1/488.
60 Top-secret telegram CRO to UK HC, 13 June 1956.
61 Top-secret telegram CRO to UK HC, 13 June 1956.
62 Top-secret telegram CRO to UK HC, 13 June 1956.
63 Top-secret telegram CRO to UK HC, 13 June 1956.
64 Admiralty, Top-secret note about Operation Mosaic, Office of D.C.N.S, 21 June 1955, presented at Executive for Mosaic ad-hoc meeting on 6 July 1955 at the Ministry of Supply. TNA: FCO 1/8.
65 David Dunbar-Nasmith, Commander HMS *Alert*, Report of Proceedings 24 March to 30 April 1956 at Monte Bello, 4 May 1956. TNA: ADM 296/45.
66 Dunbar-Nasmith, Report of Proceedings, May 1956.
67 Martell, Mosaic – Monte Bello Atomic Tests, 1956, p. 80.
68 'Nuclear Energy Seen as Main Source of Power for Shipping', *Canberra Times*, 16 April 1956, p. 1.
69 Gareth Jones, 'United Kingdom Submarine Nuclear Propulsion', *British Journal for Military History*, vol. 7, issue 3, November 2021, pp. 112–13.
70 AB Sainsbury, obituary, Vice-Admiral Sir Hugh Martell, *The Independent*, 25 February 1999, <www.independent.co.uk/arts-entertainment/obituary-viceadmiral-sir-hugh-martell-1073024>, accessed 14 January 2025.
71 Hugh Martell, Evidence to the Royal Commission, Transcript of Proceedings. NAA: A6448, 10, p. 4879.
72 Martell, Evidence to the Royal Commission, p. 4879.
73 Martell, Evidence to the Royal Commission, p. 4879.
74 Martell, Mosaic – Monte Bello Atomic Tests, 1956, p. 1.
75 Minutes of the third meeting of the Monte Bello Working Party held at 339 Swanston Street, Melbourne, on Friday 30 September 1955. NAA: A6456, R029/207.

76 Questions likely to arise after initial announcement of plans for Mosaic and Buffalo, attached to top-secret telegram from Commonwealth Relations Office to UK high commissioner in Australia (acting), 10 September 1955. TNA: ES 1/488.
77 Questions on Mosaic and Buffalo, September 1955.
78 Questions on Mosaic and Buffalo, September 1955.
79 Symonds, *A History of British Atomic Tests in Australia*, p. 336.
80 Royal Commission into British Nuclear Tests in Australia, *Report* (JR McClelland, president), 2 vols, AGPS, Canberra, 1985, p. 483.
81 Symonds, *A History of British Atomic Tests in Australia*, pp. 336–37.
82 Minutes of the sixth meeting of the Monte Bello Working Party held at 339 Swanston Street, Melbourne, on Friday 8 December 1955. NAA: A6456, R029/207.
83 Department of Supply, Draft press statement, 'Plans for tests in 1956 at Monte Bello and Maralinga', undated. NAA: A6456, R209/004.
84 AS Brown, secretary, Prime Minister's Department, Top-secret cyphered teleprinter message to Frederick Shedden, secretary, Department of Defence, 8 September 1955. NAA: A6456, R209/004.
85 Brown, Memo to Shedden, 8 September 1955.
86 Brown, Memo to Shedden, 8 September 1955.
87 'Atom Bomb Blasts Here Next Year: H-bomb Device?', *Sun* (Sydney), 13 September 1955.
88 Prime Minister's Department, Statement on nature of tests, 12 September 1955. NAA: A6456, R209/004.
89 PM's Department, Statement on nature of tests, September 1955.
90 Commonwealth Relations Office, Top-secret telegram to UK high commissioner in Australia (Acting), 29 July 1955. TNA: ES 1/488.
91 CRO, Telegram to acting UK high commissioner, July 1955.
92 Commonwealth Relations Office, Telegram to UK high commissioner in Australia (acting), 16 December 1955, R.A.F. Task Force for Atomic Weapons Test, draft official announcement. TNA: ES 1/488.
93 Minutes of the first meeting of the Monte Bello Working Party held at Navy Office, St Kilda Road, Melbourne, on Wednesday 14 September 1955. NAA: A6456, R029/207.
94 Commonwealth Relations Office, Confidential telegram to UK high commissioner in Australia (acting), 15 December 1955, draft text of proposed Mosaic announcement. TNA: ES 1/488.
95 Symonds, *A History of British Atomic Tests in Australia*, p. 326.
96 '"Keep clear" Warning for Monte Bello Atomic Weapon Test', *Central Queensland Herald Register*, 5 April 1956.
97 Symonds, *A History of British Atomic Tests in Australia*, p. 325.
98 Martell, Mosaic – Monte Bello Atomic Tests, 1956, p. 83.
99 CA Adams, scientific director, and Hugh Martell, operational commander, HMS *Narvik*, Operation Mosaic Joint Trial Orders, 2 February 1956. NAA: A6456, R120/208.
100 Adams and Martell, Operation Mosaic Joint Trial Orders, February 1956.
101 Atomic Weapons Tests Safety Committee, Media release, 'Weather Forecasting for the 1956 Monte Bello Tests', 26 June 1956. NAA: A6456, R087/088.
102 Symonds, *A History of British Atomic Tests in Australia*, p. 320.
103 AWTSC, 'Weather Forecasting', June 1956.
104 Symonds, *A History of British Atomic Tests in Australia*, p. 318.
105 Richard Fotheringham, Statement to the Royal Commission [Royal Commission into British Nuclear Tests in Australia], Statements from UK Witnesses: A–L. NAA: A6449, 1.

106 Fotheringham, Statement to the Royal Commission.
107 Richard Fotheringham, Evidence to the Royal Commission, Transcript of Proceedings. NAA: A6448, 10, p. 5487.
108 Fotheringham, Evidence to the Royal Commission, p. 5487.
109 Fotheringham, Evidence to the Royal Commission, p. 5487.
110 Fotheringham, Evidence to the Royal Commission, p. 5487.

Chapter 8: Back to the islands

1 George Halley, lieutenant, Royal Australian Navy, Report of the commander of HMAS *Motor Water Lighter 251*, 17 May 1956. NAA: A6455, AG8.
2 Lieutenant Commander Smith, Royal Australian Navy, captain of HMS *Junee*, Reports from commanding officers of *Narvik*, *Junee*, *MR1* and *Fremantle*, NAA A6455 AG8.
3 Atomic Weapons Research Establishment, Operation Mosaic – Joint Operational Plan (MJOP) – Section A11, January 1956. NAA: A6454, Z515.
4 Hugh Martell, Mosaic – Monte Bello Atomic Tests, 1956, Report by Operational Commander. NAA: A6455, RC233, p. 75.
5 Martell, Mosaic – Monte Bello Atomic Tests, 1956, p. 75.
6 Martell, Mosaic – Monte Bello Atomic Tests, 1956, p. 162.
7 Minutes of the fourth meeting of the Monte Bello Working Party held at 339 Swanston Street, Melbourne, on Friday 9 November 1955. NAA: A6456, R029/207.
8 Minutes of the fourth meeting of the Monte Bello Working Party, November 1955.
9 Atomic Weapons Research Establishment, Operation Mosaic – Joint Operational Plan (MJOP) – Section A11, January 1956. NAA: A6454, Z515.
10 AWRE, Operation Mosaic – Joint Operational Plan, January 1956.
11 AWRE, Operation Mosaic – Joint Operational Plan, January 1956.
12 Martell, Mosaic – Monte Bello Atomic Tests, 1956, p. 119.
13 Martell, Mosaic – Monte Bello Atomic Tests, 1956, p. 80.
14 Martell, Mosaic – Monte Bello Atomic Tests, 1956, p. 4
15 Martell, Mosaic – Monte Bello Atomic Tests, 1956, p. 84.
16 Martell, Mosaic – Monte Bello Atomic Tests, 1956, p. 84.
17 Martell, Mosaic – Monte Bello Atomic Tests, 1956, p. 84.
18 Martell, Mosaic – Monte Bello Atomic Tests, 1956, p. 89.
19 Martell, Mosaic – Monte Bello Atomic Tests, 1956, p. 89.
20 Howard Beale, minister for supply, Questions, House of Representatives Hansard, 6 June 1956.
21 Beale, House of Representatives Hansard, 6 June 1956.
22 Cyril Chambers, Labor member for Adelaide, Questions, House of Representatives Hansard, 6 June 1956.
23 George Halley, lieutenant, Royal Australian Navy, Report of Proceedings, May 1956, MWL 251, 11 June 1956. TNA: ADM 296/44.
24 Second progress report of Monte Bello Working Party on Operation Mosaic, 14 November 1955. NAA: A6456, R029/207.
25 George Halley, lieutenant, Royal Australian Navy, Report of the commander of the HMAS *Motor Refrigerator Lighter 252*, February 1956, sent to Commodore Special Squadron on *Narvik*, 17 March 1956. TNA: ADM 296/45.
26 George Halley, lieutenant, Royal Australian Navy, Report of the commander of HMAS *Motor Water Lighter 251*, 17 May 1956. NAA: A6455, AG8.
27 Halley, Report of Proceedings, May 1956.
28 Halley, Report of Proceedings, May 1956.

29 Halley, Report of Proceedings, May 1956.
30 Alan Butement, Atomic Weapons Tests Safety Committee, Operation Mosaic – Long range fallout sampling, 4 April 1956. NAA: A6456, R006/012.
31 Butement, AWTSC long range fallout sampling.
32 Butement, AWTSC long range fallout sampling.
33 RG Maughan, Submission tendered to the Royal Commission. NAA: A6455, RC470.
34 Maughan, Submission to the Royal Commission.
35 DA Keys, Submission tendered to the Royal Commission. NAA: A6455, RC470.
36 Keys, Submission to the Royal Commission.
37 Keys, Submission to the Royal Commission.
38 John Berry, Statement to the Royal Commission [Royal Commission into British Nuclear Tests in Australia], Statements from UK Witnesses: A–L]. NAA: A6449, 1.
39 John Berry, Evidence to the Royal Commission, Transcript of Proceedings. NAA: A6448, 9, p. 4094.
40 Berry, Statement to the Royal Commission.
41 Berry, Evidence to the Royal Commission, p. 4094.
42 Minutes of the ninth meeting of the Monte Bello Working Party held at 339 Swanston Street, Melbourne, on Friday 16 April 1956. NAA: A6456, R029/207.
43 Mosaic Executive (MOSEX), Minutes of meeting held 7 December 1955. TNA: ES 1/472.
44 MOSEX minutes, 7 December 1955.
45 Julian Cavanaugh, lieutenant commander, HMAS *Fremantle*, Hotshot, Report of Proceedings, 28 May 1956. TNA: ADM 296/42.
46 Handwritten minute note addressed to Commodore Special Squadron, attached to HMAS *Fremantle*, Hotshot, Report of Proceedings, 8 June 1956. TNA: ADM 296/42.
47 Cavanaugh, Hotshot, Report of Proceedings, May 1956.
48 Cavanaugh, Hotshot, Report of Proceedings, May 1956.
49 Halley, Report of Proceedings May 1956.
50 Cavanaugh, Hotshot, Report of Proceedings, May 1956.
51 Cavanaugh, Hotshot, Report of Proceedings, May 1956.
52 Martell, Mosaic – Monte Bello Atomic Tests, 1956, p. 143.
53 JS George, lieutenant, HMAS *Junee*, Secret note to Lieutenant Commander Smith, HMAS *Junee*, 16 May 1956. TNA: ADM 296/21.
54 Hugh Martell, Top-secret note to Lieutenant Commander Smith, HMAS *Junee*, 16 May 1956. TNA: ADM 296/21.
55 David Dunbar-Nasmith, commander HMS *Alert*, Report of Proceedings 24 March to 30 April 1956 at Monte Bello, 4 May 1956. TNA: ADM 296/45.
56 Hugh Martell, Mosaic – Monte Bello Atomic Tests, 1956. Report by Operational Commander. NAA: A6455, RC233, p. 227.
57 Martell, Mosaic Report by the Operational Commander, p. 227.
58 Dunbar-Nasmith, Report of Proceedings, May 1956.
59 Martell, Mosaic Report by the Operational Commander, p. 227.
60 Dunbar-Nasmith, Report of Proceedings, May 1956.
61 Dunbar-Nasmith, Report of Proceedings, May 1956.
62 Dunbar-Nasmith, Report of Proceedings, May 1956.
63 David Dunbar-Nasmith, commander HMS *Alert*, Report of Proceedings 1 May to 30 June 1956. TNA: ADM 296/44.
64 Dunbar-Nasmith, Report of Proceedings, 1 May to 30 June 1956.
65 JL Symonds, *A History of British Atomic Tests in Australia*, AGPS, Canberra, 1985, p. 327.

66 'Phillpot, Henry Robert 1919–2018', *Encyclopaedia of Australian Science and Innovation*, <www.eoas.info/biogs/P003294b>, accessed 29 September 2025.
67 Robert Southern, 'A Personal Recollection of Meteorological Aspects of the British Weapons Tests, Montebello Islands, Maralinga 1956–57 and Subsequent Minor Trials', statement to the Royal Commission, May 1985. NAA: A6455, RC496.
68 Southern, 'A Personal Recollection'.
69 Southern, 'A Personal Recollection'.
70 Southern, 'A Personal Recollection'.
71 Southern, 'A Personal Recollection'.
72 David Watts and George Brock, 'Ill Wind at Monte Bello', *The Times* (London), 18 June 1984.
73 Lorna Arnold and Mark Smith, *Britain, Australia and the Bomb: The Nuclear Tests and Their Aftermath*, 2nd edn, Palgrave Macmillan, New York, 2006, p. 125.
74 Martell, Mosaic – Monte Bello Atomic Tests, 1956, p. 93.
75 Ieuan Maddock, aboard HMS *Narvik*, Top-secret message to Roy Pilgrim, AWRE, 22 June 1956. TNA: ES 38/ZDRMM.
76 Martell, Mosaic – Monte Bello Atomic Tests, 1956, p. 97.
77 Martell, Mosaic – Monte Bello Atomic Tests, 1956, p. 97.
78 Martell, Mosaic – Monte Bello Atomic Tests, 1956, p. 97.
79 'Spectacular Blast On Monte Bellos', *Canberra Times*, 20 June 1956, p. 1.
80 William Penney, Transcript of Proceedings, Royal Commission into British Nuclear Tests in Australia, NAA: A6448, 12, p. 7060.
81 Stewart Stubbs, Evidence to the Royal Commission, Transcript of Proceedings. NAA: A6448, 14, p. 7588.
82 Stubbs, Evidence to the Royal Commission, p. 7591.
83 Stubbs, Evidence to the Royal Commission, p. 7593.
84 Stubbs, Evidence to the Royal Commission, p. 7591.
85 Stubbs, Evidence to the Royal Commission, p. 7591.
86 'Definitely No A-test Risk', *Sunraysia Daily* (Mildura), 26 June 1956, p. 1.
87 Royal Commission into British Nuclear Tests in Australia, *Report* (JR McClelland, president), vol 1, AGPS, Canberra, 1985, p. 250.

Chapter 9: What was G2?

1 Ernest Titterton, Evidence to the Royal Commission, Transcript of Proceedings. NAA: A6449, 14, p. 7621.
2 John McIsaac, 'Canberra Diary', *Tribune* (Sydney), 13 June 1956.
3 Zeb Leonard, 'Tampering with History: Varied Understandings of Operation Mosaic', *Journal of Australian Studies*, vol. 38, no. 2, May 2014, p. 206.
4 William Penney, Nuclear Warheads – Present Position, Top-secret memorandum Annex A, presented to the meeting of Chiefs of Staff, September 1954. TNA: DEFE 7/2208.
5 Leonard, 'Tampering with History', p. 209.
6 Leonard, 'Tampering with History', p. 209.
7 Leonard, 'Tampering with History', p. 208.
8 Leonard, 'Tampering with History', p. 208.
9 Ministry of Defence, Megaton Weapons, Notes from Working Party meeting, 14 April 1954. TNA: DEFE 7/2208.
10 Ministry of Defence, Megaton Weapons, Notes from Working Party meeting, April 1954.

11 Ministry of Defence, Megaton Weapons, Notes from Working Party meeting, April 1954
12 Sir Frederick Brundrett, UK Chiefs of Staff Committee meeting, Working Party on the Operational Use of Atomic Weapons, Notes, September 1954. TNA: DEFE 7/2208.
13 Penney, Nuclear Warheads – Present Position, September 1954.
14 Richard Moore, 'British Nuclear Warhead Design 1958–66: How Much American Help?', *Defence Studies*, vol. 4 no. 2, 2004, p. 210.
15 Penney, Nuclear Warheads – Present Position, September 1954.
16 Lorna Arnold and Mark Smith, *Britain, Australia and the Bomb: The Nuclear Tests and Their Aftermath*, 2nd edn, Palgrave Macmillan, New York, 2006, p. 106.
17 Penney, Nuclear Warheads – Present Position, September 1954.
18 John R Walker, *British Nuclear Weapons and the Test Ban 1954–1973*, Ashgate Publishing, Farnham, 2010, p. 30.
19 Walker, *British Nuclear Weapons and the Test Ban*, p. 31.
20 UK Chiefs of Staff Committee meeting, Notes, Annex B, September 1954. TNA: DEFE 7/2208.
21 Lesley Wright, Dave Wright and Nicholas Hill, 'Developing the H-Bomb: An Exercise in Calculated Decision Making', *Proceedings of the Second VALDOR Symposium Addressing Transparency in Risk Assessment and Decision Making*, Stockholm, 10–14 June 2001, p. 91.
22 Ministry of Defence, Megaton Weapons, Notes from Working Party meeting, April 1954.
23 Roger Makins (Lord Sherfield), 'William George Penney, OM, KBE, Baron Penney of East Hendred, 24 June 1909–3 March 1991', obituary, Royal Society Publishing, <royalsocietypublishing.org/doi/10.1098/rsbm.1994.0017>, accessed 25 April 2025.
24 William Penney, Thermonuclear bomb program, Top-secret note for Sir Frederick Morgan, 4 February 1955. TNA: DEFE 7/2208.
25 Penney, Nuclear Warheads – Present Position, September 1954.
26 Penney, Thermonuclear bomb program note, February 1955.
27 Penney, Thermonuclear bomb program note, February 1955.
28 Wright et al., 'Developing the H-Bomb', p. 92.
29 Penney, Thermonuclear bomb program note, February 1955.
30 Penney, Thermonuclear bomb program note, February 1955.
31 Penney, Thermonuclear bomb program note, February 1955.
32 William Penney, Letter to Sir Edwin Plowden, chair of the United Kingdom Atomic Energy Authority, about Mosaic 'Giraffe' arrangements, 1 October 1955. TNA: DEFE 16/213.
33 Penney, Letter to Plowden, 1 October 1955.
34 Steve Connor, 'The Nuclear Blast that Britain Kept Secret', *New Scientist*, 24 May 1984, p. 4.
35 'Report Sparks New Inquiries: WA Nuclear Test "Larger Than Believed"', *Canberra Times*, 26 May 1984, p. 3.
36 UK Atomic Energy Authority and AWRE report – Control of Radiation Exposure Outside the Test Area on Mosaic, undated. TNA: ADM 296/1.
37 UKAEA and AWRE report, Control of Radiation Exposure Outside the Test Area on Mosaic.
38 Charles Adams, Top-secret message to William Penney, 22 June 1956. TNA: ES 38/ZDRMM.
39 Adams, Top-secret message to Penney, June 1956.
40 Adams, Top-secret message to Penney, June 1956.

41 Roy Pilgrim, 'Yield measurements from trials', Top-secret report, Atomic Weapons Research Establishment, 14 February 1957. TNA: ES 38/ZFL7N.
42 Pilgrim, 'Yield measurements from trials'.
43 ELM Beale, Admiralty Research Laboratory, Top-secret report, 'Note on A.R.L./R.3/.791', March 1957. TNA: ES 38/ZFHGF.
44 R Potter and AC Purdie, Operation Mosaic: Air Blast Measurements, AWRE Report No. T23/57. NAA: A6455, RC241.
45 Potter and Purdie, Operation Mosaic: Air Blast Measurements.
46 Potter and Purdie, Operation Mosaic: Air Blast Measurements.
47 Potter and Purdie, Operation Mosaic: Air Blast Measurements.
48 Potter and Purdie, Operation Mosaic: Air Blast Measurements.
49 Potter and Purdie, Operation Mosaic: Air Blast Measurements.
50 Potter and Purdie, Operation Mosaic: Air Blast Measurements.
51 William Penney, Transcript of evidence to the Royal Commission into British Nuclear Tests in Australia. NAA A6448, 9, p. 4304.
52 Arnold and Smith, *Britain, Australia and the Bomb*, pp. 124–25.
53 BL White, Department of Defence, Minute paper to senior private secretary of the department, 10 January 1985. NAA: A13064, DPIE84/001705.
54 'Tsar Bomba', Cold War History, Atomic Heritage Foundation, <ahf.nuclearmuseum.org/ahf/history/tsar-bomba>, accessed 13 March 2025.
55 Commonwealth Relations Office, Top-secret telegram from to UK high commissioner in Australia (acting), 10 September 1955. TNA: ES 1/488.
56 CRO telegram to acting UK high commissioner, September 1955.
57 Arnold and Smith, *Britain, Australia and the Bomb*, p. 124.
58 CRO telegram to acting UK high commissioner, September 1955.
59 Howard Beale, minister for supply, Statement, 'No hazard from A-blast at Monte Bello', 27 May 1956. NAA: A6456, R087/088.
60 JL Symonds, *A History of British Atomic Tests in Australia*, AGPS, Canberra, 1985, p. 352.
61 Symonds, *A History of British Atomic Tests in Australia*, p. 357.
62 'Ship at Newcastle Still Radio-active', *Canberra Times*, 27 June 1956, p. 1.
63 Symonds, *A History of British Atomic Tests in Australia*, p. 350.
64 Symonds, *A History of British Atomic Tests in Australia*, p. 351.
65 David Watts and George Brock, 'Ill Wind at Monte Bello', *The Times* (London), 18 June 1984.
66 'Beale Denies Danger in Drifting Cloud', *Canberra Times*, 21 June 1956, p. 1.
67 Stuart Parker, 'A-bomb Clouds "Over WA Coast"', *Canberra Times*, 6 January 1985, p. 1.
68 Bernard Perkins, Evidence to the Royal Commission, Transcript of Proceedings. NAA: A6448, 9, p. 4167.
69 Penney, Evidence to the Royal Commission, p. 4355.
70 WG McGregor, first assistant secretary, Uranium and Nuclear Division, Department of Resources and Energy, Letter to Minister Peter Walsh, 22 June 1984. NAA: A13064 / DPIE84/001705.
71 McGregor, Letter to Peter Walsh, 22 June 1984.
72 Perkins, Evidence to the Royal Commission, p. 4171.
73 Perkins, Evidence to the Royal Commission, p. 4171.
74 Ieuan Maddock, Top-secret cable to Roy Pilgrim, AWRE, 24 June 1956. TNA: ES 38/ZDRMM.
75 Penney, Evidence to the Royal Commission, p. 4433.

76 Arthur Fadden, Statement by Acting Prime Minister Fadden, 25 June 1956, in JL Symonds, *A History of British Atomic Tests in Australia*, AGPS, Canberra, 1985, p. 363.
77 Fadden, Statement, June 1956, in Symonds, p. 364.
78 Fadden, Statement, June 1956, in Symonds, p. 364.
79 Penney, Evidence to the Royal Commission, p. 4497.
80 Roger Makins (Lord Sherfield), 'William George Penney, OM, KBE, Baron Penney of East Hendred, 24 June 1909–3 March 1991'.

Chapter 10: *Diana*, guinea pig ship

1 Captain JR Gower, 'Pierhead Jump–Some Jump!', *The RUSI Journal*, vol. 140, no. 3, 1995, pp. 51–54.
2 Susie Boniface, *Exposed: The Secret History of Britain's Nuclear Experiments*, Pen and Sword, Barnsley, 2024, p. 209.
3 Mosaic Executive (MOSEX), minutes of meeting held 7 December 1955. TNA: ES 1/472.
4 Admiralty, Top-secret report No. 4646R, 'Nuclear Explosions, Operation Mosaic. Effect of Fall-out in Machinery Spaces: Trials in HMS *Diana*', September 1956. TNA: ADM 265/90.
5 Gower, 'Pierhead Jump–Some Jump!', p. 51.
6 Gower, 'Pierhead Jump–Some Jump!', p. 52.
7 Lorna Arnold and Mark Smith, *Britain, Australia and the Bomb: The Nuclear Tests and Their Aftermath*, 2nd edn, Palgrave Macmillan, New York, 2006, p. 108.
8 Provision of a ship for fallout measurements during MOSAIC. Secret notes of a meeting held by vice chief of the naval staff (VCNS) on 9 December 1955. TNA: ADM 296/25.
9 Provision of a ship, VCNS.
10 Provision of a ship, VCNS.
11 BW Soole, Statement by the Director Physical Research, Admiralty, on Proposed Naval Trials and Observations at Operation Mosaic, 31 October 1955. TNA: ADM 296/25.
12 Soole, Admiralty statement.
13 Operation Mosaic – Fall out Observations: Procedural Details for the Stationing of H.M.S. *Diana* and the Scientific Whaler, Director Tactical, Ship Requirements and Staff Duties, Admiralty, Operation MOSAIC: stationing of HMS *Diana* for observation of fall out, 13 March 1956. TNA: ADM 1/26714.
14 Operation Mosaic – Fall out Observations: Procedural Details.
15 Operation Mosaic – Fall out Observations: Procedural Details.
16 Operation Mosaic, Firing of Weapon G.1 – Orders for H.M.S. *Diana*, May 1956.
17 United States Defense Nuclear Agency, 'Operation Castle: Summary Report of the Commander, Task Unit 13, Military Effects, Programs 1–9', 30 January 1959, Headquarters Field Command, Armed Forces Special Weapons Project, Sandia Base, Albuquerque, New Mexico, p. 3.
18 HMS *Diana*'s Report on Operation Mosaic, Military Branch, Admiralty, 30 October 1956. TNA: ADM 296/3.
19 HMS *Diana*'s Report on Operation Mosaic.
20 EML Beale, 'Operation Mosaic: The Fallout Analysed with Reference to H.M.S. *Diana*', Report for tripartite conference on effects of atomic weapons, London, September 1957. NAA: A6455, RC434.
21 Albert Matthewman, 'Operation Mosaic: Theoretical Predictions', AWRE Report No. T24/57, 15 April 1957. NAA: A6455, RC270.

22 HMS *Diana*'s Report on Operation Mosaic.
23 HMS *Diana*'s Report on Operation Mosaic.
24 Hugh Martell, Mosaic – Monte Bello Atomic Tests, 1956. Report by Operational Commander. NAA: A6455, RC233, p. 112.
25 Operation Mosaic, Firing of Weapon G.1 – Orders for H.M.S. *Diana*, Office of the Commodore, Special Squadron, HMS *Narvik* at Monte Bello, 4 May 1956. TNA: ADM 296/25.
26 Martell, Mosaic – Monte Bello Atomic Tests, 1956, p. 83.
27 Operation Mosaic, Firing of Weapon G.1 – Orders for H.M.S. *Diana*, May 1956.
28 Martell, Mosaic – Monte Bello Atomic Tests, 1956, p. 113.
29 Operation Mosaic, Firing of Weapon G.1 – Orders for H.M.S. *Diana*, May 1956.
30 John Gower, Report by Captain of HMS *Diana* on Operation Mosaic, 1956. TNA: DEFE 16/359.
31 Beale, Report for tripartite conference, September 1957.
32 Beale, Report for tripartite conference, September 1957.
33 Beale, Report for tripartite conference, September 1957.
34 Beale, Report for tripartite conference, September 1957.
35 Beale, Report for tripartite conference, September 1957.
36 Beale, Report for tripartite conference, September 1957.
37 Beale, Report for tripartite conference, September 1957.
38 Gower, Captain's report, Operation Mosaic.
39 Gower, Captain's report, Operation Mosaic.
40 Gower, Captain's report, Operation Mosaic.
41 Operation Mosaic, Firing of Weapon G.1 – Orders for H.M.S. *Diana*, May 1956.
42 Codes to be used in Interchange of Signals between C.T.F. 308 and H.M.S *Diana*, Annexe Two, Orders for HMS Diana, FLASHLIGHT FOXTROT. TNA: ADM 296/43.
43 Operation Mosaic, Firing of Weapon G.1 – Orders for H.M.S. *Diana*, May 1956.
44 Operation Mosaic, Firing of Weapon G.1 – Orders for H.M.S. *Diana*, May 1956.
45 Director Tactical, Ship Requirements and Staff Duties, Admiralty, Operation MOSAIC: Stationing of HMS Diana for observation of fall out, 13 March 1956. TNA: ADM 1/26714.
46 Head of Military Branch, DTSD, Admiralty, 24 April 1956. TNA: ADM 1/26714.
47 JA Hole, Comments on a visit to H.M.S. *Diana* to discuss protective measures, 4 May 1956. TNA: ADM 296/26.
48 Ieuan Maddock, note to Commodore Special Squadron, 4 May 1956. TNA: ADM 296/26.
49 Keith Syder, Statement to the Royal Commission [Royal Commission into British Nuclear. Tests in Australia], Statements from UK Witnesses: M–Z]. NAA: A6449, 2, p. 2.
50 Syder, Statement to the Royal Commission, p. 3.
51 Syder, Statement to the Royal Commission, p. 3.
52 Syder, Statement to the Royal Commission, p. 3.
53 Keith Syder, Evidence to the Royal Commission, Transcript of Proceedings. NAA: A6448, 9, p. 4224.
54 Syder, Statement to the Royal Commission, p. 4.
55 Syder, Evidence to the Royal Commission, p. 4225.
56 Syder, Evidence to the Royal Commission, p. 4226.
57 Martell, Mosaic – Monte Bello Atomic Tests, 1956, p. 113.
58 Martell, Mosaic – Monte Bello Atomic Tests, 1956, p. 113.

59 Kenneth Stephens, Statement to the Royal Commission [Royal Commission into British Nuclear Tests in Australia], Statements from Australian Witnesses: P–S. NAA: A6450, 4, p. 1.
60 Stephens, Statement to the Royal Commission, p. 3.
61 Stephens, Statement to the Royal Commission, pp. 3–4.
62 Stephens, Statement to the Royal Commission, p. 4.
63 Stephens, Statement to the Royal Commission, p. 5.
64 Stephens, Statement to the Royal Commission, p. 6.
65 Howard Coles, Submission tendered to the Royal Commission. NAA: A6455, RC470.
66 Coles, Submission to the Royal Commission.
67 Ronald Scott, Evidence to the Royal Commission, Transcript of Proceedings. NAA: A6448, 5, p. 2821.
68 Scott, Evidence to the Royal Commission, p. 2822.
69 Scott, Evidence to the Royal Commission, p. 2821.
70 Martell, Mosaic – Monte Bello Atomic Tests, 1956, p. 112.
71 Arthur Andrews, Evidence to the Royal Commission, Transcript of Proceedings. NAA: A6448,4, p. 1331.
72 Andrews, Evidence to the Royal Commission, p. 1332.
73 Andrews, Evidence to the Royal Commission, p. 1332.
74 Andrews, Evidence to the Royal Commission, p. 1332.
75 Symonds, *A History of British Atomic Tests in Australia,* p. 327.
76 Gower, 'Pierhead Jump–Some Jump!', p. 54.
77 Susie Boniface, 'Operation Mosaic – The First Human Experiment', Special Report, 'The Damned', *Daily Mirror* online, 4 January 2021, <https://damned.mirror.co.uk/chapter2.html>, accessed 22 September 2025.
78 'UK Atmospheric Nuclear Tests, Factsheet 5: UK Programme', <assets.publishing.service.gov.uk/media/5a755567e5274a59fa717577/ntvfactsheet5.pdf>, accessed 29 September 2025.

Chapter 11: Air operations

1 Eric McHardie, Statement to the Royal Commission [Royal Commission into British Nuclear Tests in Australia], Statements from Australian Witnesses: K–O. NAA: A6450, 3.
2 Department of Defence, 'Atomic Tests in Monte Bello Islands: Air Force Contribution', 8 June 1955. NAA: A6456, R209/005.
3 Noah Pearce, Statement to the Royal Commission [Royal Commission into British Nuclear Tests in Australia] Statements from UK Witnesses: M–Z]. NAA: A6449, 2.
4 Department of Resources and Energy, Details of RAAF Participation in British Atomic Tests Held in Australia 1952–1957, October 1983. NAA: A6456, R021/001 Part 76.
5 WL Hely, air commodore, RAAF Pearce, Letter to RAAF Headquarters Eastern Area Penrith, 21 October 1952. NAA: A6455, RC78.
6 Paul Grace, *Operation Hurricane: The Story of Britain's First Atomic Test in Australia and the Legacy that Remains*, Hachette, Sydney, 2023, p. 154.
7 Hely, Letter to RAAF Headquarters Eastern Area Penrith, October 1952.
8 Hely, Letter to RAAF Headquarters Eastern Area Penrith, October 1952.
9 No. 5 Airfield Construction Squadron, Virtual War Memorial Australia, <vwma.org.au/explore/units/486>, accessed 25 April 2025.
10 Commonwealth Relations Office, Telegram to UK high commissioner in New Zealand, 19 May 1952. TNA: FCO 1/4.

11 Royal Commission into British Nuclear Tests in Australia, *Report* (JR McClelland, president), 2 vols, AGPS, Canberra, 1985, p. 117.
12 DH Peirson, AERE, UK Atomic Energy Agency Report T113/54 for AWRE, The Results of Aerial Radiological Survey Over the Australian Coastline Between Onslow and Broome, March 1955. NAA: A6455, RC306, p. 1.
13 Royal Commission *Report*, p. 133.
14 Royal Commission *Report*, p. 133.
15 Royal Commission *Report*, p. 133.
16 Royal Commission *Report*, p. 136.
17 JM McIntyre, Counsel for the Commonwealth, Submission to the Royal Commission into British Nuclear Tests in Australia, Transcript of the Royal Commission. NAA: A6448, 17, pp. 10,119–10,120.
18 Hubert Gale, Statement to the Royal Commission [Royal Commission into British Nuclear Tests in Australia], Statements from UK Witnesses: A–L]. NAA: A6449, 1.
19 Gale, Statement to the Royal Commission.
20 Gale, Statement to the Royal Commission.
21 William Penney, Evidence to the Royal Commission, Transcript of Proceedings. NAA: A6448, 9, p. 4305.
22 Royal Commission *Report*, p. 131.
23 Royal Commission *Report*, p. 131.
24 Gale, Statement to the Royal Commission.
25 John Hampshire, Statement to the Royal Commission [Royal Commission into British Nuclear Tests in Australia], Statements from UK Witnesses: E–J]. NAA: A6450, 2.
26 Hampshire, Statement to the Royal Commission.
27 Hampshire, Statement to the Royal Commission.
28 Hampshire, Statement to the Royal Commission.
29 ID McLachlan, air commodore, air officer commanding North Eastern Area, Townsville, Operation Hurricane Order No. 4/52, 19 September 1952. NAA: A6455, RC78.
30 HJ Gale and RN Crooks, AERE, AWRE Report No.T88/54, Operation Hurricane Group Repots (Part 50), The Collection of Radioactive Cloud Samples by Aircraft Sweeps Aldermaston, September 1954. TNA: ES 5/88.
31 Gale and Crooks, AWRE Report No. T88/54, September 1954.
32 Air Vice Marshal Valston Hancock, deputy chief of the air staff, Operation Hurricane Instruction No. 5/52, 27 August 1952. NAA: A6455, RC78.
33 EO Davis, air vice marshal, Confidential message for Gale at Broome and Crooks at Townsville, 15 September 1952. NAA: A6455, AG10.
34 Noel McPhail, squadron leader, senior air staff officer, Operation Hurricane, Report by North Eastern Area Headquarters, Townsville, 20 January 1953. NAA: A6456, R021/001 Part 76.
35 McPhail, Report by North Eastern Area Headquarters, January 1953.
36 McPhail, Report by North Eastern Area Headquarters, January 1953.
37 McPhail, Report by North Eastern Area Headquarters, January 1953.
38 McPhail, Report by North Eastern Area Headquarters, January 1953.
39 McPhail, Report by North Eastern Area Headquarters, January 1953.
40 McPhail, Report by North Eastern Area Headquarters, January 1953.
41 Noel McPhail jnr, Personal correspondence with author, 4 October 2022.
42 Royal Commission *Report*, p. 130.
43 Royal Commission *Report*, p. 130.
44 Royal Commission *Report*, p. 130.

45 Royal Commission *Report*, p. 130.
46 Gale, Statement to the Royal Commission.
47 Hely, Letter to RAAF Headquarters Eastern Area Penrith, October 1952.
48 Gale, Statement to the Royal Commission.
49 Gale, Statement to the Royal Commission.
50 Gale and Crooks, AWRE Report No. T88/54, September 1954.
51 Hubert Gale, Evidence to the Royal Commission, Transcript of Proceedings. NAA: A6448, 10.
52 Gale, Evidence to the Royal Commission.
53 Peirson, Report T113/54, p. 8.
54 Peirson, Report T113/54, p. 11.
55 DH Peirson, Statement to the Royal Commission [Royal Commission into British Nuclear Tests in Australia], Statements from UK Witnesses: M–Z. NAA: A6449, 2.
56 DH Peirson, Evidence to the Royal Commission, Transcript of Proceedings. NAA: A6448, 11.
57 Peirson, Statement to the Royal Commission.
58 Peirson, Statement to the Royal Commission.
59 Peirson, Report T113/54, p. 5.
60 Peirson, Statement to the Royal Commission.
61 Peirson, Evidence to the Royal Commission, p. 5574.
62 Peirson, Evidence to the Royal Commission, p. 5568.
63 Peirson, Evidence to the Royal Commission, p. 5577.
64 Peirson, Evidence to the Royal Commission, p. 5578.
65 Peirson, Evidence to the Royal Commission, p. 5578.
66 Peirson, Evidence to the Royal Commission, p. 5593.
67 William Bovill, Statement to the Royal Commission [Royal Commission into British Nuclear Tests in Australia], Statements from Australian Witnesses: A–D. NAA: A6450, 1.
68 Bovill, Statement to the Royal Commission.
69 William Bovill, Evidence to the Royal Commission, Transcript of Proceedings. NAA: A6448, 3, p. 1001.
70 Bovill, Evidence to the Royal Commission, p. 1002.
71 Bovill, Evidence to the Royal Commission, p. 1004.
72 Robin Gray, Commander 86 (T) Wing RAAF, Letter to Headquarters, Eastern Area, RAAF, Penrith, 20 January 1953. NAA: A6455, RC78.
73 Gray, Letter to Headquarters, Eastern Area, RAAF.
74 Frank Bladin, air vice marshal and air member for personnel, Letter to air officer commanding, Eastern Area Headquarters, RAAF, Penrith, 25 March 1953. NAA: A12372, R/21923/H.
75 Stewart Menaul, Evidence to the Royal Commission, Transcript of Proceedings. NAA: A6448, 12, p. 6984.
76 Commonwealth Relations Office, Telegram to UK High Commissioner in Australia (Acting), 16 December 1955, R.A.F. Task Force for Atomic Weapons Test, draft official announcement. TNA: ES 1/488.
77 Hugh Martell, Mosaic – Monte Bello Atomic Tests, 1956. Report by Operational Commander. NAA: A6455, RC233, p. 125.
78 Martell, Mosaic – Monte Bello Atomic Tests, 1956, p. 74.
79 CRO draft official announcement, December 1955.
80 CRO draft official announcement, December 1955.
81 CRO draft official announcement, December 1955.

82 William Penney, Memorandum to Sir Frederick Brundrett, 22 December 1955, in Jordi Bigues – Archive Palomares, UAB Digital Document Repository (ddd.uab.cat), item PALO_0276, <ddd.uab.cat/pub/palomares/1955/PALO_0276_WGPenney_22des1955.pdf>, accessed 26 September 2025.
83 Penney memo to Brundrett, 22 December 1955.
84 DG Stevenson, Aircraft Decontamination, AWRE Report T33/57 Operation Mosaic, July 1957. NAA: A6455, RC287.
85 Operation Mosaic (detonations on 16 May 1956 and 19 June 1956 at Monte Bello Island, W.A.), Military Compensation SRCA Manuals and Resources Library, Liability Handbook, Ch 30 Appendices, 4.3 'Operation Mosaic', Australian Government Department of Veterans' Affairs, <clik.dva.gov.au/military-compensation-srca-manuals-and-resources-library/liability-handbook>, accessed 20 January 2025.
86 Martell, Mosaic – Monte Bello Atomic Tests, 1956, p. 123.
87 Operation Mosaic (detonations on 16 May 1956 and 19 June 1956 at Monte Bello Island, W.A.), Department of Veterans' Affairs.
88 Operation Mosaic (detonations on 16 May 1956 and 19 June 1956 at Monte Bello Island, W.A.), Department of Veterans' Affairs.
89 Minutes of the fourth meeting of the Monte Bello Working Party held at 339 Swanston Street, Melbourne, on Friday 20 October 1955. NAA: A6456, R029/207.
90 DG Stevenson, Group Leader Decontamination Group, AWRE Aldermaston. Operations Mosaic and Buffalo: The Handling, Servicing and Decontamination of Radioactive Aircraft, February 1958, AWRE Report No. T63/57. NAA: A6455, RC320.
91 Stevenson, Aircraft Decontamination, July 1957.
92 Stevenson, Aircraft Decontamination, July 1957.
93 Royal Commission *Report*, p. 425.
94 DG Stevenson, Statement to the Royal Commission [Royal Commission into British Nuclear Tests in Australia], Statements from UK Witnesses: M–Z]. NAA: A6449, 2.
95 Stevenson, Statement to the Royal Commission.
96 Stevenson, Aircraft Decontamination, July 1957.
97 Stevenson, Aircraft Decontamination, July 1957.
98 Stevenson, Aircraft Decontamination, July 1957.
99 Stevenson, Aircraft Decontamination, July 1957.
100 Stevenson, Aircraft Decontamination, July 1957.
101 Stevenson, Aircraft Decontamination, July 1957.
102 Stevenson, Aircraft Decontamination, July 1957.
103 JA Hole, Evidence to the Royal Commission, Transcript of Proceedings. NAA: A6448, 10, p. 5165.
104 Hole, Evidence to the Royal Commission, p. 5165.
105 Hole, Evidence to the Royal Commission, p. 5165.
106 Hole, Evidence to the Royal Commission, p. 5173.
107 Hole, Evidence to the Royal Commission, p. 5167.
108 Hole, Evidence to the Royal Commission, p. 5171.
109 Hole, Evidence to the Royal Commission, p. 5174.
110 'Royal Commission Hears About Monte Bello Atomic-Test: Witness Tells of Fallout on Aircraft on Mainland', *Canberra Times*, 23 February 1985, p. 12.
111 'Royal Commission hears about Monte Bello atomic test', *Canberra Times*, February 1985.
112 Matthewman, Albert, 'Operation Mosaic: Theoretical predictions', AWRE Report No. T24/57, 15 April 1957. NAA: A6455, RC270.

Epilogue: Time and tide

1 Howard Beale, Australian minister for supply, 'Why We Hold A-Tests in Australia', 6 August 1956. NAA: A6456, R047/011.
2 Report of Proceedings for HMS *Alert*, 1 May to 30 June 1956. TNA: ADM 296/44.
3 Department of Supply, Departmental note, 4 April 1957. NAA: A6456, R087/088.
4 'A-Bomb Islands Still Hot', *West Australian*, 5 April 1957.
5 'A-dust on Island Still', *Sunday Telegraph*, 31 August 1958.
6 R Rhoades, naval officer-in-charge, Letter to the director of the Commonwealth X-Ray and Radium Laboratory, 8 August 1960. NAA: A6456, R098/021.
7 Lieutenant Watson, Royal Australian Navy, Note, undated. NAA: A6456, R098/021.
8 'Radioactive Leeuwin Land Rover "A Pub Tale": Claim', *Fremantle Herald*, 27 November 2015, <heraldonlinejournal.com/2015/11/27/radioactive-leeuwin-land-rover-a-pub-tale-claim/>, accessed 30 September 2025.
9 John Moroney and Malcolm Cooper, Australian Radiation Laboratory, Environment Radiation at the Monte Bello Islands from Nuclear Weapons Tests conducted in 1952 and 1956, December 1982. NAA: A6456, R013/004.
10 State X-Ray Laboratory Preliminary Report to the Radiological Advisory Council, Radiation Survey: Monte Bello Islands, Statement and report – Visit to Trimouille Island, 29/10/74. NAA: A6455, RC485.
11 BE King and LM Davies, Appendix 1, State X-Ray Laboratory Preliminary Report to the Radiological Advisory Council, Radiation Survey: Monte Bello Islands, 19 November 1974. NAA: A6455 / RC485.
12 King and Davies, Radiation Survey, November 1974.
13 King and Davies, Radiation Survey, November 1974.
14 King and Davies, Radiation Survey, November 1974.
15 King and Davies, Radiation Survey, November 1974.
16 R Murrell and BM Hartley, State X-Ray Laboratory, Report to the Radiological Council, Second Radiation Survey: Monte Bello Islands, 27 October 1977. NAA: A6455 / RC485.
17 Murrell and Hartley, Second Radiation Survey, October 1977.
18 Murrell and Hartley, Second Radiation Survey, October 1977.
19 Murrell and Hartley, Second Radiation Survey, October 1977.
20 Murrell and Hartley, Second Radiation Survey, October 1977.
21 BM Hartley, Monte Bello Island Visit, 23rd–28th October 1978, Preliminary Report, 9 November 1978. NAA: A6455, RC485.
22 Hartley, Monte Bello Visit Report, 1978.
23 Hartley, Monte Bello Visit Report, 1978.
24 Hartley, Monte Bello Visit Report, 1978.
25 Hartley, Monte Bello Visit Report, 1978.
26 Cabinet Minute, 'Monte Bello Islands: Allegations about Radioactivity Levels and Safety Precautions', Decision No. 6897 (Ad Hoc), 17 October 1978. NAA: A13075, 6897/AD HOC.
27 ABC Education, 'Four Corners: 1950s Nuclear Tests on Monte Bello Islands', <www.abc.net.au/education/four-corners-1950s-nuclear-tests-on-monte-bello-islands/14014394>, accessed 10 March 2025.
28 Malcolm Cooper, Keith Lokan and Geoffrey Williams, 'The Radiological Status of the Monte Bello Islands; May 1983', Australian Radiation Laboratory, October 1983, p. 2.
29 Cooper et al., 'Radiological Status of the Monte Bello Islands', p. 4.
30 Royal Commission into British Nuclear Tests in Australia, *Report: Conclusions and*

Recommendations (JR McClelland, president, J Fitch and WJA Jonas, commissioners), AGPS, Canberra, 1985, p. 596.

31 Royal Commission *Report*, p. 603.

32 Expert Committee on the Review of Data on Atmospheric Fallout Arising from British Nuclear Tests in Australia, *Report* (CB Kerr, Chair) (Kerr Report), 31 May 1984, p. 34.

33 Kerr Report, p. 35.

34 Alan Clark, chair, AIRAC, appendix to letter to Barry Cohen, minister for home affairs and environment, 13 August 1984. NAA: A6456, R008/006

35 'Fitch Remains Member of N-test Royal Commission', *Canberra Times*, 18 October 1984, p. 7.

36 'Fitch Remains Member of N-test Royal Commission', *Canberra Times*, 18 October 1984.

37 James McClelland, president, Royal Commission into British Nuclear Tests in Australia, London sittings, 3 January 1985. NAA: 6448, 9, p. 4090.

38 McClelland, 3 January 1985.

39 SC Darby, GM Kendall, TP Fell, JA O'Hagan, CR Muirhead, JR Ennis, AM Ball, JA Dennis and R Doll, 'A Summary Of Mortality and Incidence of Cancer in Men from the United Kingdom Who Participated in the United Kingdom's Atmospheric Nuclear Weapon Tests and Experimental Programmes', *British Medical Journal*, vol. 296, 30 January 1988, p. 332.

40 Labrats, 'NRPB Study – A Lemon?', *Invisible Enemy*, 7 August 2025, <www.labrats.international/post/nrpb-study-a-lemon>, accessed 6 January 2026.

41 David Alton, Liberal member for Liverpool Mossley Hill, Debates, House of Commons Hansard, vol. 56, Wednesday 12 March 1984.

42 Geoffrey Pattie, minister of state for defence procurement, Debates, House of Commons Hansard, vol. 56, Wednesday 12 March 1984.

43 Gerry Kendall and Mark P Little, 'The New Study of UK Nuclear Test Veterans', *Journal of Radiological Protection*, vol. 42, no. 2, May 2022, p. 2.

44 Kendall and Little, 'The New Study of Nuclear Test Veterans', p. 6.

45 UK Cabinet, Minutes of full Cabinet meeting, 10 Downing Street, 5 December 1985. TNA: CAB 128/81.

46 UK Cabinet minutes, 5 December 1985.

47 UK Cabinet minutes, 5 December 1985.

48 Charles Powell, UK Prime Minister's Office, Briefing notes prepared for Margaret Thatcher (based on material supplied by the Foreign and Commonwealth Office), 18 April 1986. TNA: PREM 19/1675 f26.

49 'Montebello Islands Returned to WA Control', Western Australian government media statement, 8 July 1992, <www.wa.gov.au/government/media-statements/Lawrence Labor Government/Montebello-Islands-returned-to-WA-control-19920708>, accessed 27 October 2025.

50 *Western Australian Government Gazette*, No. 97, 7 July 1992, p. 3083.

51 'Montebello Islands Returned to WA Control', July 1992.

52 'Montebello Islands Returned to WA Control', July 1992.

53 Patsy Richards, Nuclear test veterans – compensation, House of Commons Library, Standard Note SNSC-05145, 31 January 2013.

54 Richards, Nuclear test veterans – compensation, January 2025.

55 Nuclear Test Veterans Legal Action Update, 23 May 2025, <www.labrats.international/post/nuclear-test-veterans-legal-action-update-23rd-may-2025>, accessed 23 September 2025.

56 Al Carns, minister for veterans and people, 'Nuclear Test Veteran Records Exercise and the Merlin Database', Debates, House of Commons Hansard, vol. 770, 30 June 2025.
57 Rebecca Long Bailey, 'Nuclear Test Veterans: Medical Records', Debates, House of Commons Hansard, vol. 741, 28 November 2023.
58 Long Bailey, House of Commons Hansard, November 2023.
59 Susie Boniface, *Exposed: The Secret History of Britain's Nuclear Experiments*, Pen and Sword, Barnsley, 2024, p. vi.
60 Johnny Mercer, minister for veterans' affairs, 'Nuclear Test Veterans: Medals', House of Commons Hansard, vol. 723, Tuesday 22 November 2022.
61 David Alton, Liberal member for Liverpool Mossley Hill, Debates, House of Commons Hansard, vol. 70, Wednesday 19 December 1984.
62 Mathew P Johansen, David P Child, Michael AC Hotchkis, Andrea Johansen, Sangeeth Thiruvotha, and Scott D Whiting, 'Radionuclides in Sea Turtles at the Montebello Islands Former Nuclear Test Sites: Current and Historical Dose Rates for Adults and Embryos', *Marine Pollution Bulletin*, vol. 158 (2020), p. 1.
63 Johansen et al., 'Radionuclides in Sea Turtles at the Montebello Islands', p. 1.
64 Johansen et al., 'Radionuclides in Sea Turtles at the Montebello Islands', p. 1.
65 Johansen et al., 'Radionuclides in Sea Turtles at the Montebello Islands', p. 4.
66 MP Johansen, DP Child, T Cresswell, JJ Harrison, MAC Hotchkis, NR Howell, A Johansen, S Sdraulig, S Thiruvoth, E Young and SD Whiting, 'Plutonium and Other Radionuclides Persist Across Marine-to-Terrestrial Ecotopes in the Montebello Islands Sixty Years after Nuclear Tests', *Science of the Total Environment*, vol. 691, 2019, p. 572.
67 Johansen et al., 'Plutonium and Other Radionuclides Persist', p. 580.
68 Madison Williams-Hoffman, Megan Cook, Jack K Clegg, Ross Kleinschmidt, Pere Masque, and Mathew P Johansen, 'Investigation of the Distribution of Transuranic Radionuclides in Marine Sediment at the Montebello Islands, Western Australia', *Journal of Environmental Radioactivity*, vol. 278, 2024, article 107505, p. 1.
69 Williams-Hoffman et al., 'Investigation of the Distribution of Transuranic Radionuclides', p. 6.
70 Williams-Hoffman et al., 'Investigation of the Distribution of Transuranic Radionuclidess', p. 1
71 Ieuan Maddock, Top-secret cable to Roy Pilgrim, AWRE, 24 June 1956. TNA: ES 38/ZDRMM.

BIBLIOGRAPHY

NAA: National Archives of Australia
TNA: The National Archives (UK)

Media sources (author named)

Bray, TC, 'And Monte Bello is Still "Hot"', *Courier-Mail* (Brisbane), 6 August 1953, p. 1.
Connor, Steve, 'The Nuclear Blast that Britain Kept Secret', *New Scientist*, 24 May 1984, p. 4.
Daily News Special Correspondent (Jack Coulter), 'Pressmen find Montebellos like Wartime Beachhead', *Daily News* (Perth), 29 April 1952, p. 2.
McIsaac, John 'Canberra Diary', *Tribune* (Sydney), 13 June 1956.
Parker, Stuart, 'A-bomb Clouds "over WA Coast"', *Canberra Times*, 6 January 1985, p. 1.
Pincher, Chapman, 'Atom Bomb Wizard', *Western Mail* (Perth), 6 March 1952, p. 7.
Stephens, Sean-Paul, and Wally, Ian, 'Poignant Tribute to Dorrie: Memories of Pilbara Massacre, Nuclear Bombs and the Last Marthudunera Speaker', *Pilbara News*, 9 September 2025.
Titterton, Ernest, 'Why Australia Is Preferred as an Atom Testing Site', *Sydney Morning Herald*, 16 May 1956.
Warner, Denis, 'British Atomic Test Explosion Successful: Step Towards Hydrogen Bomb: "Trigger" Experiment', *Daily Telegraph*, 17 May 1956.
Watts, David, and Brock, George, 'Ill Wind at Monte Bello', *The Times* (London), 18 June 1984.

Media sources (author unnamed) chronological order

'Likely Site of British Bomb Experiment', *West Australian*, 22 April 1952, p. 1.
'Possible Site of Atomic Tests', *West Australian*, 22 April 1952, p. 10.
'Atomic Fleet Sails from Fremantle', *West Australian*, 23 April 1952, p. 1.
'Churchill in Control of A-bomb Test', *News* (Adelaide), 15 May 1952, p. 1.
'2200-Mile Mission to Cover Atom Bomb Story', *The Journalist*, July 1952, p. 2.
'Man Who Hates Fuss to Blast Atom Bomb', *Daily Mirror* (Sydney), 15 July 1952, p. 10.
'Security at Atom Islands Worries Our Navy Chiefs', *Melbourne Herald*, 2 August 1952, p. 3.
'Britain Tests Her First Atomic Bomb off W.A. Coast', *West Australian*, 4 October 1952, p. 1.
'Newsmen Kept a Long Watch on a Hilltop', *West Australian*, 4 October 1952, p. 11.
'Atomic History', *West Australian,* 4 October 1952, p. 2.
'Pressmen's Long Watch on Monte Bello', *The Age* (Melbourne), 4 October 1952, p. 2.
'Britain Joins the Atomic Powers', *The Age*, 6 October 1952, p. 2.
'"Space suit" Scientists Check on Atom Blast', *Sun News-Pictorial*, Melbourne, 6 October 1952, p. 3.
'Dr. Penney Knighted', *Telegraph* (Brisbane), 24 October 1952, p. 1.

'Brisbane Airman Wins Medal at Cocos Island', *Courier-Mail* (Brisbane), 3 November 1952, p. 1.

'Sir W Penney to Broadcast on Atomic Blast', *Canberra Times*, 4 November 1952, p. 1.

'After Cocos the R.A.A.F. Turns to Townsville, *Townsville Daily Bulletin*, 18 December 1952, p. 6.

Editorial, *Daily Express*, 15 October 1953, p. 4.

'Monte Bello Expedition Reaches Onslow', *West Australian*, 9 November 1953, p. 8.

'Atom Bomb Blasts Here Next Year: H-bomb Device?', *Sun* (Sydney), 13 September 1955.

' "Keep Clear" Warning for Monte Bello Atomic Weapon Test', *Central Queensland Herald Register*, 5 April 1956.

'Nuclear Energy Seen as Main Source of Power for Shipping', *Canberra Times*, 16 April 1956, p. 1.

'Protests Flood in as H-hour Nears: Stop Bomb Tests They Say', *Tribune* (Sydney), 25 April 1956.

'Spectacular Blast on Monte Bellos', *Canberra Times*, 20 June 1956, p. 1.

'Threat to 3 Towns: Atom Cloud Drift', *Sun* (Sydney), 21 June 1956.

'A-cloud Adrift – But No Danger', *The Argus* (Melbourne), 21 June 1956, p. 1.

'Definitely No A-test Risk', *Sunraysia Daily* (Mildura), 26 June 1956, p. 1.

'Beale Denies Danger in Drifting Cloud', *Canberra Times*, 21 June 1956, p. 1.

'Labor Will Stop Weapons Test', *Canberra Times*, 25 June 1956.

'Safe in London ...', *Tribune* (Sydney), 27 June 1956.

'Ship at Newcastle Still Radio-active', *Canberra Times*, 27 June 1956, p. 1.

'A-Bomb Islands Still Hot', *West Australian*, 5 April 1957.

'A-dust on Island Still', *Sunday Telegraph*, 31 August 1958.

'Report Sparks New Inquiries: WA Nuclear Test "Larger Than Believed"', *Canberra Times*, 26 May 1984, p. 3.

'Fitch Remains Member of N-test Royal Commission', *Canberra Times*, 18 October 1984, p. 7.

'Nuclear Test Inquiry: Radioactive Waste "Harmless"', *Canberra Times*, 23 January 1985, p. 12.

'Royal Commission Hears about Monte Bello Atomic-test: Witness Tells of Fallout on Aircraft on Mainland', *Canberra Times*, 23 February 1985, p. 12.

Primary sources

Adams, CA, Top-secret letter to Dr DH Black, UKMOSS Melbourne, 29 September 1955. TNA: ADM 296/1.

Adams, CA, scientific director, and Martell, Hugh, operational commander, HMS *Narvik*, Operation Mosaic Joint Trial Orders, 2 February 1956. NAA: A6456, R120/208.

Admiralty, Top-secret note about Operation Mosaic, Office of D.C.N.S, 21 June 1955, presented at Executive for Mosaic ad-hoc meeting on 6 July 1955 at the Ministry of Supply. TNA: FCO 1/8.

Admiralty, Top-secret report No. 4646R, 'Nuclear Explosions, Operation Mosaic. Effect of Fall-out in Machinery Spaces: Trials in HMS *Diana*', September 1956. TNA: ADM 265/90.

Alton, David, Liberal member for Liverpool Mossley Hill, Debates, House of Commons Hansard, vol. 56, Wednesday 12 March 1984.

Alton, David, Liberal member for Liverpool Mossley Hill, Debates, House of Commons Hansard, vol. 70, Wednesday 19 December 1984.

Andrews, Arthur, Evidence to the Royal Commission, Transcript of Proceedings. NAA: A6448,4, p. 1319.

Andrews, Arthur, Royal Australian Navy service record. NAA: A6769.
Atomic Weapons Research Establishment, Operation Mosaic – Joint Operational Plan (MJOP) – Section A11, January 1956. NAA: A6454, Z515.
Atomic Weapons Tests Safety Committee, Media release, 'Weather Forecasting for the 1956 Monte Bello Tests', 26 June 1956. NAA: A6456, R087/088.
Atomic Weapons Tests Safety Committee, Minutes of Fourth Meeting, University of Melbourne, 10 January 1956. NAA: A6456, R120/137 PART 1.
Atomic Weapons Tests Safety Committee, Minutes of Eighth Meeting, University of Melbourne, 23 July 1956. NAA: A6456, R120/137 PART 1.
Attlee, Clement, Top-secret and personal cable to Robert Menzies, 16 September 1950. NAA: A6456, R084/001.
Attlee, Clement, UK prime minister, Telegram to Robert Menzies, Australian prime minister, 21 September 1950. TNA: FCO 1/1.
Bailey, Sir Kenneth, solicitor-general, Secret letter to AS Brown, Prime Minister's Department, 18 July 1952. NAA: A816, 3/301/534.
Barnes, DE, Fall Out of Contamination for the Monte-Bello Explosion: Criteria for Safe Firing, AWRE report T1/53, May 1953. Some preliminary results from the Monte Bello tests, relevant to defence, including civil defence. NAA: A6455, RC604.
Beale, EML, 'Operation Mosaic: The Fallout Analysed with Reference to H.M.S. *Diana*', Report for tripartite conference on effects of atomic weapons, London, September 1957. NAA: A6455, RC434.
Beale, Howard, minister for supply, Statement, 'No Hazard from A-blast at Monte Bello', 27 May 1956. NAA: A6456, R087/088.
Beale, Howard, minister for supply, Questions, House of Representatives Hansard, 6 June 1956.
Beale, Howard, minister for supply, 'Why We Hold A-Tests in Australia', 6 August 1956. NAA: A6456, R047/011.
Berry, John, Evidence to the Royal Commission, Transcript of Proceedings. NAA: A6448, 9, p. 4094.
Berry, John, Statement to the Royal Commission [Royal Commission into British Nuclear Tests in Australia], Statements from UK Witnesses: A–L]. NAA: A6449, 1.
Bevin, Ernest, secretary of state for foreign affairs, Answer to question from Mr Blackburn, House of Commons Hansard, vol. 472, Wednesday 22 March 1950.
Blackburn, Albert, Question to the secretary of state for foreign affairs, House of Commons Hansard, vol. 472, Wednesday 22 March 1950.
Bladin, Frank, air vice marshal and air member for personnel, Letter to air officer commanding, Eastern Area Headquarters, RAAF, Penrith, 25 March 1953. NAA: A12372, R/21923/H.
Bourke, WM, Questions, House of Representatives Hansard, 29 August 1952.
Bovill, William, Evidence to the Royal Commission, Transcript of Proceedings. NAA: A6448, 3
Bovill, William, Statement to the Royal Commission [Royal Commission into British Nuclear Tests in Australia], Statements from Australian Witnesses: A–D. NAA: A6450, 1.
Brook, Norman, Cabinet secretary, Clandestine Use of Atomic Weapons, Top-secret memorandum to Clement Attlee, prime minister, 12 July 1951. TNA: PREM 11/560.
Brooking, PW, Note to E Evans-Lombe, HUREX, attached to Leonard Tyte, Operation Hurricane: Dr Tyte's Diary, 1952. TNA: ES 1/78.
Brooking, PW, Note attached to PF Cooper, Operation Hurricane: Report by Capt. PF Cooper RN (Retd), 4 October 1954. TNA: ES 1/147.

Brown, AS, secretary, Prime Minister's Department, Secret reply to Ben Cockram, office of the UK high commissioner, 9 July 1952. NAA: NAA: A816, 3/301/534.

Brown, AS, secretary, Prime Minister's Department, Top-secret memo to Frederick Shedden, secretary, Department of Defence, 23 September 1952. NAA: A6456, R095/004.

Brown, AS, secretary, Prime Minister's Department, Secret letter to the Admiralty, 9 November 1954. TNA: AVIA 65/1151.

Brown, AS, secretary, Prime Minister's Department, Top-secret cyphered teleprinter message to Frederick Shedden, secretary, Department of Defence, 8 September 1955. NAA: A6456, R209/004.

Brown, AS, secretary, Prime Minister's Department, Top-secret memo to the secretary of the Department of Supply, 13 September 1955. NAA: A6456, R029/207.

Brundrett, Sir Frederick, Ministry of Defence, Top-secret letter to PH Dean, Foreign Office, 13 July 1954. TNA: DEFE 7/897.

Brundrett, Sir Frederick, UK Chiefs of Staff Committee meeting, Working Party on the Operational Use of Atomic Weapons, notes, September 1954. TNA: DEFE 7/2208.

Brundrett, Sir Frederick, Working Party on the Operational Use of Atomic Weapons, Memorandum by the Chairman, presented to the meeting of Chiefs of Staff, September 1954. TNA: DEFE 7/2208.

Bunting, EJ, acting secretary to Cabinet, Secret Cabinet minute, Decision No. 431, Submission No. 272 – Atomic Weapon Test – Security, 27 May 1952. NAA: A816, 3/301/534.

Butement, Alan, Atomic Weapons Tests Safety Committee, Operation Mosaic – Long range fallout sampling, 4 April 1956. NAA: A6456, R006/012.

Cabinet APEX Committee, Minutes of first meeting, No. 10 Downing Street, 26 March 1952. TNA: CAB 134/734.

Cabinet APEX Committee, Minutes of second meeting, Cabinet Office, 15 April 1952. TNA: CAB 134/734.

Cabinet APEX Committee, Minutes of third meeting No. 10 Downing Street, 24 April 1952. TNA: CAB 134/734.

Cabinet Minute, 'Monte Bello Islands: Allegations about Radioactivity Levels and Safety Precautions', Decision No. 6897 (Ad Hoc), 17 October 1978. NAA: A13075, 6897/AD HOC.

Carns, Al, minister for veterans and people, 'Nuclear Test Veteran Records Exercise and the Merlin Database', Debates, House of Commons Hansard, vol. 770, Monday 30 June 2025.

Carter, Henry, Evidence to the Royal Commission, Transcript of Proceedings. NAA: A6448, 10.

Carter, Henry, Statement to the Royal Commission [Royal Commission into British Nuclear Tests in Australia], Statements from UK Witnesses: A–L]. NAA: A6449, 1.

Cavanaugh, Julian, Lieutenant Commander, HMAS *Fremantle*, c 28 May 1956. TNA: ADM 296/42.

Chambers, Cyril, Labor member for Adelaide, Questions, House of Representatives Hansard, 6 June 1956.

Churchill, Winston, prime minister, House of Commons Hansard, vol. 505, Thursday 23 October 1952.

Churchill, Winston, Top-secret Cabinet document 14 December 1954: 'Notes on Tube Alloys, 1954'. DSND 14/4 Churchill Archives Centre.

Churchill, Winston, prime minister, Statement on Defence, House of Commons Hansard, vol. 537, Tuesday 1 March 1955.

Clark, Alan, chair, AIRAC, Letter to Barry Cohen, minister for home affairs and environment, 13 August 1984. NAA: A6456, R008/006.
Clarke, EJS, office of the paymaster-general, Letter to the CRO, 17 September 1952. TNA: FCO 1/5.
Cockcroft, Sir John, Handwritten note to William Penney, 6 September 1952. TNA: ES 1/498.
Cockram, Ben, office of the UK high commissioner, Secret letter to Allan McKnight, 13 June 1952. NAA: A816, 3/301/534.
Cockram, Ben, office of the UK high commissioner, Secret letter to Allan McKnight, 1 July 1952. NAA: A816, 3/301/534.
Codes to be used in Interchange of Signals between C.T.F. 308 and H.M.S *Diana*, Annexe Two, Orders for HMS *Diana*, FLASHLIGHT FOXTROT. TNA: ADM 296/43.
Coles, Howard, Submission tendered to the Royal Commission. NAA: A6455, RC470.
Collins, Vice Admiral Sir John, UKSLS, Top-secret letter to chief of the naval staff, 21 August 1951, Operation Hurricane, copies of correspondence on operational matters between the UK Hurricane Executive and Hurricane Panel (Aust). NAA: A6456 R095/002.
Colville, Captain George, commanding HMS *Zeebrugge* senior officer of the 4th Landing Ship Squadron, Report of Proceedings, Appendix II, 'Weather', 6 June 1952. NAA: A13070, SMITH 168.
Commonwealth Relations Office, Top-secret telegram to high commissioner for the UK, 23 February 1952. NAA: A816, 3/301/534.
Commonwealth Relations Office, Telegram to UK high commissioner in New Zealand, 19 May 1952. TNA: FCO 1/4.
Commonwealth Relations Office, Top-secret telegram to UK high commissioner in Australia (Acting), 29 July 1955. TNA: ES 1/488.
Commonwealth Relations Office, Top-secret telegram from to UK high commissioner in Australia (acting), 10 September 1955. TNA: ES 1/488.
Commonwealth Relations Office, Confidential telegram to UK High Commissioner in Australia (Acting), 15 December 1955, draft text of proposed Mosaic announcement. TNA: ES 1/488.
Commonwealth Relations Office, Telegram to UK high commissioner in Australia (acting), 16 December 1955, R.A.F. Task Force for Atomic Weapons Test, draft official announcement. TNA: ES 1/488.
Commonwealth Relations Office, Top-secret telegram to UK High Commission in Australia, 13 June 1956. TNA: ES 1/488.
Cook, William, Atomic Weapons Research Establishment, Top-secret letter to Sir Richard Powell, Ministry of Defence, 25 April 1955. TNA: ES 38/ZFGWL.
Cooper, PF, Operation Hurricane: Report by Capt. PF Cooper RN (Retd), 4 October 1954. TNA: ES 1/147.
Davey, George, British High Commission Canberra, Secret letter to Allan McKnight, deputy secretary, Prime Minister's Department, 27 August 1952. NAA: A6456, R096/006.
Davey, T Division, Home Office, Civil Defence Department, Training: Memorandum on effects of the hydrogen bomb, 20 August 1954. TNA: HO 338/23.
Davis, EO, Air Vice Marshal, Confidential message for Gale at Broome and Crooks at Townsville, 15 September 1952. NAA: A6455, AG10.
Dean, PH, Foreign Office, Top-secret letter to Sir Frederick Brundrett, Ministry of Defence, 8 July 1945. TNA: DEFE 7/897.
Defence Committee, Minutes of meeting, 28 August 1952. NAA: A6456, R096/006.

Defence Committee, Minutes of meeting, 4 September 1952. NAA: A6456, R096/006.
Department of Defence, For Press: Atomic Weapon Test (media release), 13 August 1952. NAA: A816, 10/301/129.
Department of Defence minute for Minister Philip McBride, 8 May 1952. NAA: A816, 3/301/534.
Department of Defence, 'Atomic Tests in Monte Bello Islands: Air Force Contribution', 8 June 1955. NAA: A6456, R209/005.
Department of the Navy, Secret minute paper, 18 December 1952. NAA: A6456, R015/011.
Department of Resources and Energy, Details of RAAF Participation in British Atomic Tests Held in Australia 1952–1957, October 1983. NAA: A6456, R021/001 Part 76.
Department of Supply, Draft press statement 'Plans for tests in 1956 at Monte Bello and Maralinga', undated. NAA: A6456, R209/004.
Department of Supply, Departmental note, 4 April 1957. NAA: A6456, R087/088.
Dhenin, Geoffrey, Medical Officer's Journal, HMS *Tracker*. NAA: A13070, Smith 167, WE Smith vs Commonwealth of Australia Document 167 (Folder 13).
Director Tactical, Ship Requirements and Staff Duties, Admiralty, Operation MOSAIC: stationing of HMS *Diana* for observation of fall out, 13 March 1956. TNA: ADM 1/26714.
Douglas-Home, Alec, secretary of state for Commonwealth affairs, personal telegram to Anthony Eden, prime minister, 4 December 1956. TNA: CAB 21/3148.
Douglas-Home, Alec, secretary of state for Commonwealth affairs, Draft telegram to the prime minister of Australia and the prime minister of New Zealand, 12 December 1956. TNA: CAB 21/3148.
Draft Press Notice, 19 February 1951, annotated by Winston Churchill, CHUR 2/28, Churchill Archives Centre, University of Cambridge.
Dunbar-Nasmith, David, commander HMS *Alert*, Report of Proceedings 24 March to 30 April 1956 at Monte Bello, 4 May 1956. TNA: ADM 296/45.
Dunbar-Nasmith, David, commander HMS *Alert*, Report of Proceedings 1 May to 30 June 1956. TNA: ADM 296/44.
Eden, Anthony, UK prime minister, Personal note to Robert Menzies, Australian prime minister, 16 May 1955. NAA: A6456, R084/001.
Eden, Anthony, UK prime minister, Note to Robert Menzies, Australian prime minister, 8 July1955. NAA: A6456, R084/001.
Eden, Anthony, prime minister, Debate, Atom and Hydrogen Bomb Experiments, House of Commons Hansard, vol. 547, Tuesday 6 December 1955.
Eden, Anthony, prime minister, Statement to the House of Commons, Hansard, vol. 552, 16 May 1956.
Eden, Anthony, prime minister, Debates, House of Commons Hansard, vol. 554, Tuesday 19 June 1956.
Elliot, Ministry of Defence, London, Top-secret cypher telegram to Lord Tedder, BJSM Washington, 14 August 1950. TNA: FCO 1/1.
Evatt, HV, leader of the opposition, and Beale, Howard, minister for supply, Questions, House of Representatives Hansard, 15 February 1956.
Evatt, HV, leader of the opposition, Questions, House of Representatives Hansard, 21 June 1956.
Expert Committee on the Review of Data on Atmospheric Fallout Arising from British Nuclear Tests in Australia, *Report* (CB Kerr, Chair) (Kerr Report), 31 May 1984.
Fadden, AW, acting prime minister, Questions, House of Representatives Hansard, 21 June 1956.

Fadden, Arthur, Statement by Acting Prime Minister Fadden, 25 June 1956, in JL Symonds, *A History of British Atomic Tests in Australia*, AGPS, Canberra, 1985.

Fotheringham, Richard, Evidence to the Royal Commission, Transcript of Proceedings. NAA: A6448, 10.

Fotheringham, Richard, Statement to the Royal Commission [Royal Commission into British Nuclear Tests in Australia], Statements from UK Witnesses: A–L]. NAA: A6449, 1.

Gale, Hubert, Evidence to the Royal Commission, Transcript of Proceedings. NAA: A6448, 10.

Gale, Hubert, Statement to the Royal Commission [Royal Commission into British Nuclear Tests in Australia], Statements from UK Witnesses: A–L]. NAA: A6449, 1.

Gale, HJ, and Crooks, RN, AERE, AWRE Report No.T88/54, Operation Hurricane Group Repots (Part 50), The Collection of Radioactive Cloud Samples by Aircraft Sweeps Aldermaston, September 1954. TNA: ES 5/88.

Garbett, Reginald, Submission tendered to the Royal Commission. NAA: A6455, RC470.

George, JS, lieutenant, HMAS *Junee*, secret note to Lieutenant Commander Smith, HMAS Junee, 16 May 1956. TNA: ADM 296/21.

Gower, John, Report by captain of HMS *Diana* on Operation Mosaic, 1956. TNA: DEFE 16/359.

Gray, Robin, commander 86 (T) Wing RAAF, Letter to Headquarters, Eastern Area, RAAF, Penrith, 20 January 1953. NAA: A6455, RC78.

Gray, Robin, Letter to RAAF Richmond, 30 March 1953. NAA: A12372, R/21923/H.

Halley, George, lieutenant, Royal Australian Navy, Report of the commander of the HMAS *Motor Refrigerator Lighter 252*, February 1956, sent to Commodore Special Squadron on *Narvik*, 17 March 1956. TNA: ADM 296/45.

Halley, George, lieutenant, Royal Australian Navy, Report of the commander of HMAS *Motor Water Lighter 251*, 17 May 1956. NAA: A6455, AG8.

Halley, George, lieutenant, Royal Australian Navy, Report of Proceedings May 1956, MWL 251, 11 June 1956. TNA: ADM 296/44.

Hampshire, John, Statement to the Royal Commission [Royal Commission into British Nuclear Tests in Australia], Statements from UK Witnesses: E–J]. NAA: A6450, 2.

Hancock, Air Vice Marshal Valston, deputy chief of the air staff, Operation Hurricane Instruction No. 5/52, 27 August 1952. NAA: A6455, RC78.

Handwritten minute note addressed to Commodore Special Squadron, attached to HMAS *Fremantle*, Hotshot, Report of Proceedings, 8 June 1956. TNA: ADM 296/42.

Hartley, BM, Monte Bello Island Visit, 23rd–28th October 1978, Preliminary Report, 9 November 1978. NAA: A6455, RC485.

Hartley, BM, Health Department, Report of meeting at the Western Australian Department of Conservation and Environment, 17 March 1981. NAA: A6455, RC485.

Hely, WL, Air Commodore, RAAF Pearce, Letter to RAAF Headquarters Eastern Area Penrith, 21 October 1952. NAA: A6455, RC78.

Hicks, EP, AWRE Aldermaston, Minute paper to Admiral Brooking, 14 October 1954. TNA: ES 1/792.

HMS *Diana*'s Report on Operation Mosaic, Military Branch, Admiralty, 30 October 1956. TNA: ADM 296/3.

Hole, JA, Comments on a visit to H.M.S. *Diana* to discuss protective measures, 4 May 1956. TNA: ADM 296/26.

Hole, JA, Evidence to the Royal Commission, Transcript of Proceedings. NAA: A6448, 10.

Hole, JA, Statement to the Royal Commission [Royal Commission into British Nuclear Tests in Australia], Statements from UK Witnesses: A–L]. NAA: A6449, 1.

Holmes, Stephen, British High Commissioner to Australia, Top-secret telegram to Alec Douglas-Home, secretary of state for Commonwealth relations, 16 March 1955. TNA: FCO 1/8.

How, FC, Ministry of Supply, Top-secret letter to RM Saner, Foreign Office, 19 August 1952. TNA: FCO 1/4.

Hughes, Emrys, House of Commons Hansard, vol. 554, Monday 11 June 1956.

HUREX, via Admiralty, Top-secret message to Commander Task Force 4 (Torlesse), Order HTO 262, 19 September 1952. TNA: FCO 1/5.

HUREX Signal No. 1, via Admiralty to UKSLS Melbourne, 25 June 1951. TNA: DEFE 16/505.

Hurricane Panel, Remarks on UK High Commission letter of 1 July, undated. NAA: A816, 3/301/534.

Keys, DA, Submission tendered to the Royal Commission. NAA: A6455, RC470.

Kimber, Gurth, UK High Commission in Australia, Secret memo to Frank Lloyd, 28 July 1955. TNA: FCO 1/8.

King, BE, and Davies, LM, Appendix 1, State X-Ray Laboratory Preliminary Report to the Radiological Advisory Council, Radiation Survey: Monte Bello Islands, 19 November 1974. NAA: A6455 / RC485.

Lindemann, Frederick, Note to Churchill about the British position on the atomic bomb, 19 January 1950. Lindemann/Cherwell papers J97, Nuffield Library, Oxford.

Lindemann, Frederick, Note to Churchill, 29 December 1952. Lindemann/Cherwell papers, J122, Nuffield College Library, Oxford.

Lindemann, Frederick, Note 1 to Churchill, 1 December 1953. Lindemann/Cherwell papers J97, Cherwell 1950, Nuffield College Library, Oxford.

Lindemann, Frederick, Note 2 to Churchill, 1 December 1953. Lindemann/Cherwell papers J97, Nuffield College Library, Oxford.

Lloyd, Captain FB, Secret note to Vice Admiral EM Evans-Lombe, deputy chief of naval staff, Admiralty, 31 October 1952. NAA: A6455, AG10.

Lloyd, Captain FB, Secret note to Neil Pritchard, Commonwealth Relations Office, 2 January 1953. NAA: A6455, AG10.

Lloyd, Captain FB, Ministry of Supply, London, Top-secret letter to Neil Pritchard, Commonwealth Relations Office, 7 December 1953, TNA: FCO 1/6.

Lloyd, Captain FB, Secret memorandum to William Penney, 7 January 1955. TNA: AVIA 65/1151.

Long Bailey, Rebecca, Nuclear Test Veterans: Medical Records, House of Commons debate, 28 November 2023. UK Parliament Hansard, vol. 741.

Mabbutt, Graham, Statement to the Royal Commission [Royal Commission into British Nuclear Tests in Australia], Statements from UK Witnesses: M–Z. NAA: A6449, 2.

Maddock, Ieuan, Note to Commodore Special Squadron, 4 May 1956. TNA: ADM 296/26.

Maddock, Ieuan, aboard HMS *Narvik*, Top-secret message to Roy Pilgrim, AWRE, 22 June 1956. TNA: ES 38/ZDRMM.

Makins, Roger, Foreign Office, Top-secret record of conversation with Admiral Lewis Strauss, 10 March 1955. TNA: DEFE 7/897.

Maralinga Sub-Committee, Top-secret report on discussions with United Kingdom officials concerning the support desired from Australia in connection with Operation Mosaic, 1955. NAA: A6456, R031/007.

Martell, Hugh, Evidence to the Royal Commission, Transcript of Proceedings. NAA: A6448, 10.

Martell, Hugh, Mosaic – Monte Bello Atomic Tests, 1956. Report by Operational Commander. NAA: A6455, RC233.
Martell, Hugh, Top-secret note to Lieutenant Commander Smith, HMAS *Junee*, 16 May 1956. TNA: ADM 296/21.
Matthewman, Albert, 'Operation Mosaic: Theoretical predictions', AWRE Report No. T24/57, 15 April 1957. NAA: A6455, RC270.
Maudling, Reginald, minister of supply, House of Commons Hansard, vol. 554, Monday 11 June 1956.
Maughan, RG, Submission tendered to the Royal Commission. NAA: A6455, RC470.
McBride, Philip, Minister for Defence, Questions, House of Representatives Hansard, 29 August 1952.
McBride Philip, Minister for Defence, Statement, 'Australian Warships and Scientists Present at the Atomic Test', 3 October 1952. TNA: FCO 1/5.
McBride, Philip, Minister for Defence, 'Atomic Tests', Top-secret letter to Howard Beale, Minister for Supply, 8 June 1955. NAA: A6456, R209/005.
McBride, Philip, Minister for Defence, Top-secret letter to Josiah Francis, Minister for the Navy, 20 July 1955. NAA: A6456, R031/007.
McClelland, James, President, Royal Commission into British Nuclear Tests in Australia, London sittings, 3 January 1985. NAA: 6448, 9.
McEnhill, JJ, Story of Operation Hurricane, Presented 26/7/85 at Sydney, Royal Commission, NAA: A6455, RC603.
McGregor, WG, first assistant secretary, Uranium and Nuclear Division, Department of Resources and Energy, Letter to Minister Peter Walsh, 22 June 1984. NAA: A13064 / DPIE84/001705.
McHardie, Eric, Statement to the Royal Commission [Royal Commission into British Nuclear Tests in Australia], Statements from Australian Witnesses: K–O. NAA: A6450, 3.
McIntyre, JM, Counsel for the Commonwealth, Submission to the Royal Commission into British Nuclear Tests in Australia, Transcript of the Royal Commission. NAA: A6448.
McLachlan, ID, Air Commodore, air officer commanding North Eastern Area, Townsville, Operation Hurricane Order No. 4/52, 19 September 1952. NAA: A6455, RC78.
McNicoll, Alan, deputy chief of naval staff, draft advice to UK High Commissioner, 28 May 1952. NAA: A816, 3/301/534.
McPhail, Noel, squadron leader, Senior Air Staff Officer, Operation Hurricane, Report by North Eastern Area Headquarters, Townsville, 20 January 1953. NAA: A6456, R021/001 Part 76.
McPhail, Noel (jnr), Personal correspondence with author, 4 October 2022.
Menaul, Stewart, Evidence to the Royal Commission, Transcript of Proceedings. NAA: A6448, 12.
Menaul, Stewart, Statement to the Royal Commission [Royal Commission into British Nuclear Tests in Australia], Statements from UK Witnesses: M–Z. NAA: A6449, 2.
Menzies, Robert, prime minister, Statement, 'Atomic Weapon Test', 3 October 1952. TNA: FCO 1/5.
Menzies, Robert, Australian prime minister, letter to Anthony Eden, UK prime minister, 20 June 1955. NAA: A6456, R084/001.
Mercer, Johnny, minister for veterans' affairs, 'Nuclear Test Veterans: Medals', House of Commons Hansard, vol. 723, Tuesday 22 November 2022.
Milne, Norm, Handwritten notes, 17 April 1952. Milne 'Atomic Testing' file, courtesy of Rory Milne.

Milne, Norm, Memorandum to news editor, 3 June 1952, Milne 'Atomic Testing' file, courtesy of Rory Milne.

Ministry of Defence, Megaton Weapons, notes from Working Party meeting, 14 April 1954. TNA: DEFE 7/2208.

Ministry of Defence briefing note, Nuclear Tests, 18 July 1955. TNA: DEFE 7/897.

Minutes of the first meeting of the Monte Bello Working Party held at Navy Office, St Kilda Road, Melbourne, on Wednesday 14 September 1955. NAA: A6456, R029/207.

Minutes of the third meeting of the Monte Bello Working Party held at 339 Swanston Street, Melbourne, on Friday 30 September 1955. NAA: A6456, R029/207.

Minutes of the fourth meeting of the Monte Bello Working Party held at 339 Swanston Street, Melbourne, on Friday 20 October 1955. NAA: A6456, R029/207.

Minutes of the fifth meeting of the Monte Bello Working Party held at 339 Swanston Street, Melbourne, on Friday 9 November 1955. NAA: A6456, R029/207.

Minutes of the sixth meeting of the Monte Bello Working Party held at 339 Swanston Street, Melbourne, on Friday 8 December 1955. NAA: A6456, R029/207.

Minutes of the seventh meeting of the Monte Bello Working Party held at 339 Swanston Street, Melbourne, on Friday 18 January 1956. NAA: A6456, R029/207.

Minutes of the ninth meeting of the Monte Bello Working Party held at 339 Swanston Street, Melbourne, on Friday 16 April 1956. NAA: A6456, R029/207.

Montgomery, I, Ministry of Defence, letter to Frank How, Ministry of Supply, 8 March 1950. TNA: DEFE 7/2208.

Morgan, Sir Frederick, Secret letter to William Penney, 31 December 1952. TNA: ES 38/ ZDRS6.

Moroney, John, and Cooper, Malcolm, Australian Radiation Laboratory, Environment Radiation at the Monte Bello Islands from Nuclear Weapons Tests conducted in 1952 and 1956, December 1982. NAA: A6456, R013/004.

Mortimer, R, Medical Officer's Journal, HMS *Tracker*. NAA: A13070, Smith 167, WE Smith vs Commonwealth of Australia Document 167 (Folder 13).

Mosaic Executive (MOSEX), minutes of first meeting, Giles Court London, 6 July 1955. TNA: ADM 296/5.

Mosaic Executive (MOSEX), minutes of meeting held 7 December 1955. TNA: ES 1/472.

Mosaic Mission to Australia – Summary of Actions, presented to MOSEX Committee, London, 17 August 1955. TNA: ADM 296/12.

Mount Potter – Western Australia – Post Office history. NAA: C3629, MOUNT POTTER.

Murrell, R, and Hartley, BM, State X-Ray Laboratory, Report to the Radiological Council, Second Radiation Survey: Monte Bello Islands, 27 October 1977. NAA: A6455 / RC485.

Naval officer-in-charge West Australia Area, Department of Navy, Secret cable to HMAS *Hawkesbury*, December 1952. NAA: A6456, R015/011.

Naval officer-in-charge West Australian Area, Confidential note to the secretary, Naval Board, Melbourne 4 December 1952. NAA: A6456, R015/011.

Naval officer-in-charge West Australian Area, 'Valuable scientific stores recovered from the Monte Bello Area', Letter to the secretary, Naval Board, Naval Office Melbourne, 20 March 1953. NAA: A6456, R015/011.

O'Connor, Frank, secretary, Department of Supply, Top-secret briefing note for Howard Beale, minister for supply, 26 May 1955. NAA: A6456, R209/005.

O'Connor, Frank, Letter to Brigadier CCF Spry, director-general of security, 28 November 1956. NAA: A6456, R087/059.

Office of the UK High Commissioner, Secret note to Allen Brown, Prime Minister's Department Canberra, 22 October 1952. TNA: FCO 1/5.

Operation Mosaic – Fall out Observations: Procedural Details for the Stationing of H.M.S. Diana and the Scientific Whaler, Director Tactical, Ship Requirements and Staff Duties, Admiralty, Operation MOSAIC: stationing of HMS *Diana* for observation of fall out, 13 March 1956. TNA: ADM 1/26714.

Operation Mosaic, Firing of Weapon G.1 – Orders for H.M.S. *Diana*, Office of the Commodore, Special Squadron, HMS *Narvik* at Monte Bello, 4 May 1956. TNA: ADM 296/25.

Pattie, Geoffrey, Minister of State for Defence Procurement, Debates, House of Commons Hansard, vol. 56, Wednesday 12 March 1984.

Pearce, Noah, Statement to the Royal Commission [Royal Commission into British Nuclear Tests in Australia], Statements from UK Witnesses: M–Z]. NAA: A6449, 2.

Peck, EK, Evidence to the Royal Commission, Transcript of Proceedings. NAA: A6450, 4.

Peck, EK, Statement to the Royal Commission [Royal Commission into British Nuclear Tests in Australia], Statements from Australian Witnesses: P–S. NAA: A6450, 4.

Peirson, DH, AERE, UK Atomic Energy Agency Report T113/54 for AWRE, The Results of Aerial Radiological Survey Over the Australian Coastline Between Onslow and Broome, March 1955. NAA: A6455, RC306.

Peirson, DH, Evidence to the Royal Commission, Transcript of Proceedings. NAA: A6448, 11.

Peirson, DH, Statement to the Royal Commission [Royal Commission into British Nuclear Tests in Australia], Statements from UK Witnesses: M–Z. NAA: A6449, 2.

Penney, William, Epicure. Report on Proposed Site, 12 January 1951. NAA: A6455, BV1.

Penney, William, Secret note to Frederick Lindemann (Lord Cherwell), 27 February 1952. TNA: ES 1/498.

Penney, William, Top-secret note to Leslie Martin, 14 July 1952. TNA: ES 1/498.

Penney, William, Memorandum on safety standards for Operation Hurricane, 30 July 1952. NAA: A6455, RC78.

Penney, William, Letter to Frederick Lindemann (Lord Cherwell), 15 August 1952. NAA: A6455, RC468.

Penney, William, Top-secret message to Leslie Martin, 22 August 1952. TNA: FCO 1/4.

Penney, William, Final Draft of Dr Penney's Broadcast, October 1952. TNA: FCO 1/5.

Penney, William, Operation Hurricane – the dose-rate contours of the residual radioactive contamination, top-secret section of Director's Report, 27 August 1954. TNA: ES5/3.

Penney, William, Nuclear Warheads – Present Position, Top-secret memorandum Annex A, presented to the meeting of Chiefs of Staff, September 1954. TNA: DEFE 7/2208.

Penney, William, Thermonuclear bomb program, Top-secret note for Sir Frederick Morgan, 4 February 1955. TNA: DEFE 7/2208.

Penney, William, Letter to Sir Edwin Plowden, chair of the United Kingdom Atomic Energy Authority, about Mosaic 'Giraffe' arrangements, 1 October 1955. TNA: DEFE 16/213.

Penney, William, Evidence to the Royal Commission, Transcript of Proceedings. NAA: A6448, 9, and A6449, 2.

Penney, William, Transcript of Proceedings, Royal Commission into British Nuclear Tests in Australia, NAA: A6448, 12.

Perkins, Bernard, Evidence to the Royal Commission, Transcript of Proceedings. NAA: A6448,9, p. 4167.

Potter, R, and Purdie, AC, Operation Mosaic: Air Blast Measurements, AWRE Report No. T23/57. NAA: A6455, RC241.

Powell, Charles, UK Prime Minister's Office, Briefing notes prepared for Margaret Thatcher (based on material supplied by the Foreign and Commonwealth Office), 18 April 1986. TNA: PREM 19/1675 f26.

Press Reaction to Atomic Trials, Briefing document prepared by Australian Department of Supply, 1956 (exact date unknown). NAA: A6456, R047/011.

Prime Minister's Department, Statement on nature of tests, 12 September 1955. NAA: A6456, R209/004.

Pritchard, Neil, Commonwealth Relations Office, Top-secret letter to RM Saner, 19 August 1952. TNA: FCO 1/4.

Provision of a ship for fall-out measurements during MOSAIC. Secret notes of a meeting held by Vice Chief of the Naval Staff (VCNS) on 9 December 1955. TNA: ADM 296/25.

Questions likely to arise after initial announcement of plans for Mosaic and Buffalo, attached to top-secret telegram from Commonwealth Relations Office to UK high commissioner in Australia (acting), 10 September 1955. TNA: ES 1/488.

Report of Proceedings for HMS *Alert*, 1 May to 30 June 1956. TNA: ADM 296/44.

Report of the Montebello Working Party on Operation Mosaic, 10 October 1955. Prepared for the full Maralinga Committee. NAA: A6456, R029/207.

Rhoades, R, naval officer-in-charge West Australian Area, 'Alleged exposure to radioactive material from jeep used at Monte Bello', Letter to the Medical Superintendent, Repatriation General Hospital, Western Australia, 8 August 1960. NAA: A6456, R098/021.

Royal Commission into British Nuclear Tests in Australia, *Report: Conclusions and Recommendations* (JR McClelland, president, J Fitch and WJA Jonas, commissioners), AGPS, Canberra, 1985.

Royal Commission into British Nuclear Tests in Australia, *Report* (JR McClelland, president), 2 vols, AGPS, Canberra, 1985.

Salisbury, Lord, Top-secret letter to Field Marshal Earl Alexander of Tunis, 20 August 1954. TNA: DEFE 7/897.

Sandys, Duncan, Defence: Outline of Future Policy (Defence White Paper), March 1957, TNA, CAB/129/86.

Scott, Ronald, Evidence to the Royal Commission, Transcript of Proceedings. NAA: A6448, 5.

Scrivenor, Robert, Lieutenant, HMAS Hawkesbury, confidential message to the naval officer-in-charge West Australian Area, 30 November 1952. NAA: A6456, R015/011.

Second progress report of Monte Bello Working Party on Operation Mosaic, 14 November 1955. NAA: A6456, R029/207.

Secretary of state for Commonwealth relations, Top-secret telegram to high commissioner for the UK, Australia, 2 April 1955, TNA, FCO/1/8.

Serpell, David, undersecretary of the Treasury, Draft letter to be sent to HL Jenkyns, Treasury representative at the British High Commission in Australia, 13 July 1955, TNA FCO 1/8.

Silverman, Sydney, Question to the prime minister, House of Commons Hansard, vol. 505, Thursday 23 October 1952.

Smith, lieutenant commander, Royal Australian Navy, captain of HMAS *Junee*, Reports from commanding officers of *Narvik*, *Junee*, MR1 and *Fremantle*, NAA A6455 AG8.

Soole, BW, Statement by the Director Physical Research, Admiralty, on Proposed Naval Trials and Observations at Operation Mosaic, 31 October 1955. TNA: ADM 296/25.

Southern, Robert, 'A Personal Recollection of Meteorological Aspects of the British Weapons Tests, Montebello Islands, Maralinga 1956–57 and Subsequent Minor Trials', Statement to the Royal Commission, May 1985. NAA: A6455, RC496.

State X-Ray Laboratory Preliminary Report to the Radiological Advisory Council, Radiation Survey: Monte Bello Islands, Statement and report – Visit to Trimouille Island, 29/10/74. NAA: A6455, RC485.

Stephens, Kenneth, Statement to the Royal Commission [Royal Commission into British Nuclear Tests in Australia], Statements from Australian Witnesses: P–S. NAA: A6450, 4.

Stephens, Michael, Statement to the Royal Commission [Royal Commission into British Nuclear Tests in Australia], Statements from UK Witnesses: M–Z. NAA: A6449, 2.

Stevens, Donald, Director, Commonwealth X-Ray and Radium Laboratory, Letter to the secretary, Department of the Navy, 17 May 1960. NAA: A6456, R098/021.

Stevenson, DG, Aircraft Decontamination, AWRE Report T33/57 Operation Mosaic, July 1957. NAA: A6455, RC287.

Stevenson, DG, Group Leader Decontamination Group, AWRE Aldermaston. Operations Mosaic and Buffalo: The Handling, Servicing and Decontamination of Radioactive Aircraft, February 1958, AWRE Report No. T63/57. NAA: A6455, RC320.

Stevenson, DG, Statement to the Royal Commission [Royal Commission into British Nuclear Tests in Australia], Statements from UK Witnesses: M–Z. NAA: A6449, 2.

Stubbs, Stewart, Evidence to the Royal Commission, Transcript of Proceedings. NAA: A6448, 14.

Swingler, Stephen, Debates, House of Commons Hansard, vol. 554, Tuesday 19 June 1956.

Syder, Keith, Evidence to the Royal Commission, Transcript of Proceedings. NAA: A6448, 9.

Syder, Keith, Statement to the Royal Commission [Royal Commission into British Nuclear Tests in Australia], Statements from UK Witnesses: M–Z. NAA: A6449, 2.

Sykes, EL, Commonwealth Relations Office, Secret letter to Frank How, Ministry of Supply. TNA: ES 1/498.

Tedder, Lord, BJSM Washington, Top-secret cypher telegram to Ministry of Defence London, 27 July 1950. TNA: FCO 1/1.

Tedder, Lord, BJSM Washington, Top-secret cypher telegram to Ministry of Defence London, 17 October 1950. TNA: FCO 1/1.

Telegram from Commonwealth Relations Office Secretary London (draft), 20 September 1950. TNA: FCO 1/1.

Testing of United Kingdom Atomic Weapon, Top-secret ministerial minute paper prepared for the UK Prime Minister Clement Attlee, 14 September 1950. TNA: FCO 1/1.

The Rand Corporation, 'Worldwide Effects of Atomic Weapons: Project Sunshine', ASTIA Document Number AD 112389, 6 August 1953, amended 1956, Santa Monica, California.

Thomas, Andrew, Statement and Weekly Orders issue number 1053, 1/11/1954 and other orders and correspondence, Royal Commission into British Nuclear Tests in Australia, Submission to the Royal Commission, NAA: A6455, RC553.

Titterton, Ernest, Evidence to the Royal Commission, Transcript of Proceedings. NAA: A6449, 14, p. 7617.

Tory, GW, UK deputy high commissioner Canberra, Minutes of meeting between Adams, Martell, Tory and Rouse, 28 July 1956. NAA: A6455, RC267.

Tyte, Leonard, Operation Hurricane: Dr Tyte's Diary, 1952. TNA: ES 1/78.

UK Admiralty, Secret message to UKSLS Melbourne, 20 June 1951, Operation Hurricane, copies of correspondence on operational matters between the UK Hurricane Executive and Hurricane Panel (Aust). NAA: A6456 R095/002.

UK Atomic Energy Authority and AWRE report – Control of Radiation Exposure Outside the Test Area on Mosaic, undated. TNA: ADM 296/1.
UK Cabinet, Minutes of full Cabinet meeting, 10 Downing Street, 5 December 1985. TNA: CAB 128/81.
UK Chiefs of Staff Committee meeting, Notes, Annex B, September 1954. TNA: DEFE 7/2208.
UK high commissioner in Australia, Telegram to secretary of state for commonwealth relations, 30 May 1952. TNA: FCO 1/4.
UK high commissioner in Australia, Confidential telegram to Commonwealth Relations Office, 21 June 1956. TNA: ES 1/489.
UK high commissioner in Australia (acting), Secret telegram to CRO, 28 August 1952. TNA: FCO 1/5.
UKSLS, letter to Ben Cockram, 24 September 1952. Telegram (HUREX 115) to UKSLS: Martin. TNA: FCO 1/5/11.
UK Services Liaison Staff Australia, Top-secret cypher telegram to Ministry of Defence London, 13 October 1950. TNA: FCO 1/1.
UK Services Liaison Staff Australia, Operation Hurricane timeline, 17 July 1952. TNA: FCO 1/4.
United States Defense Nuclear Agency, 'Operation Castle: Summary Report of the Commander, Task Unit 13, Military Effects, Programs 1–9', 30 January 1959, Headquarters Field Command, Armed Forces Special Weapons Project, Sandia Base, Albuquerque, New Mexico.
Walkling, Alec, Evidence to the Royal Commission, Transcript of Proceedings. NAA: A6448, 9.
Walkling, Alec, Operation Hurricane: Report of the Radiological Hazard Division, (draft), Fort Halstead, 15 May 1953. NAA: A6455, RC250.
Walkling, Alec, Statement to the Royal Commission [Royal Commission into British Nuclear Tests in Australia], Statements from UK Witnesses: M–Z. NAA: A6449, 2.
Walley, Dorrie, Oral history, Burrup Stories Project, March 2010. Creator: Kate Gregory. Australian Institute of Aboriginal and Torres Strait Islander Studies, GREGORY_KOO1.
Watson, lieutenant, Royal Australian Navy, Note, undated. NAA: A6456, R098/021.
Western Australian Government Gazette, No. 97, 7 July 1992.
White, BL, Department of Defence, Minute paper to senior private secretary of the Department, 10 January 1985. NAA: A13064, DPIE84/001705.
Williams, EC, Top-secret note for Air Defence Committee Working Party, 'Some speculations about the effects of developments in atomic weapons upon future Defence policy, with a hypothesis about the consequences to the provision of air defence for the United Kingdom', 8 May 1954. TNA: DEFE 7/2208.
Wills, HA, AWTSC secretary, Memorandum to Leonard Dwyer, Commonwealth Bureau of Meteorology, 2 May 1956. NAA: A6456, R120/137 PART 1.
Wilson, Harold, leader of the opposition, Debates, House of Commons Hansard, vol. 687, Thursday 16 January 1964.

Secondary sources

Anderson, Ross, and Archer, Theresa, 'Trial and Errors', *The Great Circle*, Australian Association for Maritime History, vol. 44, no. 2, 2022, pp. 67–73.
Arnold, Lorna, 'Lord Penney', *Proceedings of the American Philosophical Society*, vol. 136, no. 3, September 1992, pp. 444–48.
Arnold, Lorna and Smith, Mark, *Britain, Australia and the Bomb: The Nuclear Tests and Their Aftermath*, 2nd edn, Palgrave Macmillan, New York, 2006.

Baylis, John, 'The Development of Britain's Thermonuclear Capability 1954–61: Myth or Reality?', *Contemporary Record*, vol. 8, no. 1, Summer 1994.

Beale, Howard, *This Inch of Time: Memoirs of Politics and Diplomacy*, Melbourne University Press, Melbourne, 1977.

Bell, Mark S, 'Beyond Emboldenment: How Acquiring Nuclear Weapons Can Change Foreign Policy', *International Security*, vol. 40, no. 1, Summer 2015, pp. 87–119.

Boniface, Susie, *Exposed: The Secret History of Britain's Nuclear Experiments*, Pen and Sword, Barnsley, 2024.

Boyle, Peter G, 'Britain, America and the Transition from Economic to Military Assistance, 1948–51', *Journal of Contemporary History*, July 1987, vol. 22, no. 3, July 1987, p. 525.

Browning, Tas, 'HMAS Hawkesbury', *Sabretache*, vol. LI, no. 1, March 2010.

Burbidge, Andrew A, Blyth, John D, Fuller, Phillip J, Kendrick, Peter G, Stanley, Frances J and Smith, Laurie A, 'The Terrestrial Vertebrate Fauna of the Montebello Islands, Western Australia', *CALMScience*, vol. 3, no. 2, pp. 95–107, January 2000.

Cooper, Malcolm, Lokan, Keith, and Williams, Geoffrey, 'The Radiological Status of the Monte Bello Islands; May 1983', Australian Radiation Laboratory, October 1983.

Darby, SC, Kendall, GM, Fell, TP, O'Hagan, JA, Muirhead, CR, Ennis, JR, Ball, AM, Dennis, JA, and Doll, R, 'A Summary of Mortality and Incidence of Cancer in Men from the United Kingdom who Participated in the United Kingdom's Atmospheric Nuclear Weapon Tests and Experimental Programmes', *British Medical Journal*, vol. 296, 30 January 1988.

Drewe, Robert, *Montebello: A Memoir*, Hamish Hamilton, Melbourne, 2012.

Dylan, Huw, 'Operation TIGRESS: Deception for Counterintelligence and Britain's 1952 Atomic Test', *Journal of Intelligence History*, vol. 14, no. 1, pp. 1–15, 14 August 2014.

Gledhill, Jim, 'The Bow and Arrow Versus the Atom Bomb: Air Defence in Scotland 1945–1955', *British Journal for Military History*, vol. 10, issue 1, March 2024.

Gott, Richard, 'The Evolution of the Independent British Deterrent', *International Affairs* (Royal Institute of International Affairs), vol. 39, no. 2, April 1963.

Gower, Captain JR, 'Pierhead Jump–Some Jump!', *The RUSI Journal*, vol. 140, no. 3, pp. 51–54, 1995.

Gowing, Margaret, with Arnold, Lorna, *Independence and Deterrence: Britain and Atomic Energy, 1945–1952,* United Kingdom Atomic Energy Authority, St Martin's Press, New York, 1974.

Gowing, Margaret, 'Monte Bello', in *Independence and Deterrence: Britain and Atomic Energy, 1945–1952,* Palgrave Macmillan, London, 1974.

Grace, Paul, *Operation Hurricane: The Story of Britain's First Atomic Test in Australia and the Legacy that Remains*, Hachette, Sydney, 2023.

Hill, Frank L, 'Notes on the Natural History of the Monte Bello Islands', *Proceedings of the Linnean Society of London,* vol. 165, issue 2, June 1955.

Hordern, Marsden, 'Australia's Debt to Stokes and the BEAGLE', *Maritime Studies*, 1992, p. 64.

Hordern, Marsden, 'Darwin's Ship "Beagle" and Her Captains', *Journal of Australian Naval History*, vol. 7 no. 1, 2010.

Hughes, Helen, 'Film and the British Atomic Project', *Historical Journal of Film, Radio and Television,* vol. 42, no. 2, 2022, pp. 219–43.

Johansen, MP, Child, DP, Cresswell, T, Harrison, JJ, Hotchkis, MAC, Howell, NR, Johansen, A, Sdraulig, S, Thiruvoth, S, Young, E, and Whiting, SD, 'Plutonium and other Radionuclides Persist Across Marine-to-terrestrial Ecotopes in the Montebello Islands Sixty Years after Nuclear Tests', *Science of the Total Environment*, vol. 691, 2019, pp. 572–83.

Johansen, Mathew P, Child, David P, Hotchkis, Michael AC, Johansen, Andrea, Thiruvotha, Sangeeth, and Whiting, Scott D, 'Radionuclides in Sea Turtles at the Montebello Islands Former Nuclear Test Sites: Current and Historical Dose Rates for Adults and Embryos', *Marine Pollution Bulletin*, vol. 158, 2020.

Jones, Gareth, 'United Kingdom Submarine Nuclear Propulsion', *British Journal for Military History*, vol. 7, issue 3, November 2021.

Kendall, Gerry, and Little, Mark P, 'The New Study of UK Nuclear Test Veterans', *Journal of Radiological Protection*, vol. 42, no. 2, May 2022.

Leonard, Zeb, 'Tampering with History: Varied Understandings of Operation Mosaic', *Journal of Australian Studies*, vol. 38, no. 2, May 2014, pp. 205–19.

Lord Hennessy of Nympsfield, 'Cabinets and the Bomb', The Inaugural Michael Quinlan Lecture, Queen's Robing Room, House of Lords, Wednesday 2 February 2011.

McKnight, David, *Australia's Spies and Their Secrets*, Allen & Unwin, Sydney, 1994.

Melissen, Jan, 'Prelude to Interdependence: The Anglo-American Relationship and the Limits of Great Britain's Nuclear Policy, 1952–1957', *Contemporary Security Policy*, vol. 11, no. 3, 1990, pp. 205–31.

Menaul, Stewart, 'Great Britain and NATO Theater Nuclear Forces', *Strategic Review*, United States Strategic Institute, Washington DC, Winter 1981.

Milliken, Robert, *No Conceivable Injury: The Story of Britain and Australia's Atomic Cover-Up*, Penguin, Ringwood Vic., 1986.

Montague, PD, 'The Monte Bello Islands', *The Geographical Journal*, vol. 42, no. 1, July 1913, pp. 34–44.

Moore, Richard, 'British Nuclear Warhead Design 1958–66: How Much American Help?', *Defence Studies*, vol. 4, no. 2, 2004, pp. 207–28.

Morris, KD, Management Proposals for the Montebello Islands and Surrounding Waters, CALM Occasional Paper No. 3/91, August 1991, Department of Conservation and Land Management, Como, Western Australia.

Norris, Robert Standish, 'Questions on the British H-Bomb', Natural Resources Defense Council, 22 June 1992.

Pelling, Henry, 'The H-Bomb and Eden's Diplomacy', in: *Churchill's Peacetime Ministry, 1951–55*, Palgrave Macmillan, London, 1997.

Perry, PF, and Wells, FE (eds), 'Survey of the Marine Fauna of the Montebello Islands, Western Australia and Christmas Island, Indian Ocean', *Records of the Western Australian Museum*, Supplement no. 59, Western Australian Museum, November 2000.

Piper, Cassia J, and Veth, Peter M, 'Palaeoecology and Sea Level Changes: Decline of Mammal Species Richness During Late Quaternary Island Formation in the Montebello Islands, North-Western Australia', *Palaeontologia Electronica*, vol 24, no. 2, a20, 2021.

Purdon, James, 'The Meaning of Monte Bello', in Beck, John, and Bishop, Ryan, *Cold War Legacies: Legacy, Theory, Aesthetics*, Edinburgh University Press, Edinburgh, 2016.

Pyne, Katherine, 'Art or Article? The Need for and Nature of the British Hydrogen Bomb, 1954–58', *Contemporary British History*, vol. 9, no. 3, 1995, pp. 562–85.

Rawles, Simon, director, 'Britain's Nuclear Bomb Scandal: Our Story', Hardcash Productions, 2024.

Reynolds, David, 'A "Special Relationship"? America, Britain and the International Order Since the Second World War', *International Affairs*, vol. 62, no.1, Winter, 1985–1986, pp. 1–20.

Reynolds, Wayne, 'Rethinking the Joint Project: Australia's Bid for Nuclear Weapons, 1945–1960', *Historical Journal*, vol. 41, no. 3, 1998.

Richards, Patsy, Nuclear Test Veterans – Compensation, House of Commons Library, Standard Note SNSC-05145, 31 January 2013.

Ruane, Kevin, *Churchill and the Bomb in War and Cold War*, Bloomsbury Academic, London, 2016.

Sherratt, Tim, 'A Political Inconvenience: Australian Scientists at the British Atomic Weapons Tests, 1952–53', *Historical Records of Australian Science*, vol. 6, no. 2, December 1985.

Stanbury, Myra, 'Mother-of-Pearl Shell Cultivation: An Early 20th Century Experiment in the Montebello Islands', *The Great Circle*, vol. 16, no. 2, 1994, pp. 90–120.

Symonds, JL, *A History of British Atomic Tests in Australia*, AGPS, Canberra, 1985.

'The Scientific Civil Service: Reorganisation and Recruitment during the Reconstruction Period', *Aircraft Engineering*, October 1945, pp. 280–88.

Todd, Richard, host, *After the Bombs* (alternatively, *After the A Bomb*), SBS documentary production, 1970. State Library of Western Australia collection.

Toohey, Brian, *Secret: The Making of Australia's Security State*, Melbourne University Press, Melbourne, 2019.

Veth, Peter, Aplin, Ken, Wallis, Lynley, Manne, Tina, Pulsford, Tim, White, Elizabeth, and Chappell, Alan, *The Archaeology of Montebello Islands, North-West Australia: Late Quaternary Foragers on an Arid Coastline*, Oxford, Archaeopress, 2007.

Walker, John R, *British Nuclear Weapons and the Test Ban 1954–1973*, Ashgate Publishing, Farnham, 2010.

Ward, I, Veth, P, and Manne, T, 'To the Islands Born: The Research Potential of Submerged Landscapes and Human Habitation Sites from the Islands of NW Australia', in Harff, J, Bailey, G, and Lüth, F (eds), *Geology and Archaeology: Submerged Landscapes of the Continental Shelf*, Geological Society, London, Special Publications, vol. 411, 2016, pp. 251–63.

Williams-Hoffman, Madison, Cook, Megan, Clegg, Jack K, Kleinschmidt, Ross, Masque, Pere, and Johansen, Mathew P, 'Investigation of the Distribution of Transuranic Radionuclides in Marine Sediment at the Montebello Islands, Western Australia', *Journal of Environmental Radioactivity*, vol. 278, 2024, article 107505.

Wilson, David, *Always First: The RAAF Construction Squadrons 1942–1974*, Air Power Study Centre, Canberra, 1998.

Wright, Lesley, Wright, Dave, and Hill, Nicholas, 'Developing the H-Bomb: An Exercise in Calculated Decision Making', *Proceedings of the Second VALDOR Symposium Addressing Transparency in Risk Assessment and Decision Making*, Stockholm, 10–14 June 2001.

Websites

ABC Education, *Four Corners*: '1950s Nuclear Tests on Monte Bello Islands', <www.abc.net.au/education/four-corners-1950s-nuclear-tests-on-monte-bello-islands/14014394>, accessed 10 March 2025.

Boniface, Susie, 'Operation Mosaic – The First Human Experiment', Special Report, The Damned, *Daily Mirror* online, 4 January 2021, <https://damned.mirror.co.uk/chapter2.html>, accessed 22 September 2025.

Dombey, Norman, 'John Clive Ward, 1 August 1924–6 May 2000', obituary, Biographical Memoirs of the Fellows of the Royal Society, 17 February 2021, <royalsocietypublishing.org/doi/10.1098/rsbm.2020.0023>, accessed 27 March 2025.

Dombey, Norman, and Grove, Eric, 'Britain's Thermonuclear Bluff', *London Review of Books*, vol. 14, no. 20, 22 October 1992, <www.lrb.co.uk/the-paper/v14/n20/norman-dombey/britain-s-thermonuclear-bluff>, accessed 27 March 2025.

Duckworth, WE, 'Ieuan Maddock: 29 March 1917–29 December 1988', Biographical Memoirs of Fellows of the Royal Society, 1 November 1991, <royalsocietypublishing.org/doi/abs/10.1098/rsbm.1991.0016>, accessed 13 January 2025.

Farrell, Darcy, 'Douglas J Burton', Doug Burton Tribute, WA TV History, 11 November 2014, <watvhistory.com/2014/11/doug-burton-tribute-part-5-of-6>, accessed 26 March 2025.

Lambrides, Ariana BJ, Ditchfield, Kane, Ulm, Sean, and Veth, Peter. 'A Moving Feast: Australia's Coastal and Island Archaeological Record', in Scott M Fitzpatrick and Jon Erlandson (eds), *The Oxford Handbook of Island and Coastal Archaeology* (online edn, Oxford Academic, 16 August 2023), <doi.org/10.1093/oxfordhb/9780197607770.013.23>, accessed 28 January 2025.

Makins, Roger (Lord Sherfield), 'William George Penney, OM, KBE, Baron Penney of East Hendred, 24 June 1909–3 March 1991', obituary, Royal Society Publishing, <royalsocietypublishing.org/doi/10.1098/rsbm.1994.0017>, accessed 25 April 2025.

Mardie Station, <inherit.dplh.wa.gov.au/public/inventory/printsinglerecord/2498a13e-726f-41f6-81cc-a06b0a916681>, accessed 23 June 2025.

'Montebello Islands Returned to WA control', Western Australian government media statement, 8 July 1992, <www.wa.gov.au/government/media-statements/Lawrence Labor Government/Montebello-Islands-returned-to-WA-control-19920708>, accessed 27 October 2025.

Newton, JO, and Jenkin, John, 'Sir Ernest William Titterton 1916–1990', *Australian Dictionary of Biography*, <adb.anu.edu.au/biography/titterton-sir-ernest-william-973>, accessed 25 September 2025.

No. 5 Airfield Construction Squadron, Virtual War Memorial Australia, <vwma.org.au/explore/units/486>, accessed 25 April 2025.

'Nuclear Radiation', The Atomic Archive, <www.atomicarchive.com/science/effects/radiation.html>, accessed 7 January 2026.

Nuclear Test Veterans Legal Action Update, 23 May 2025, <www.labrats.international/post/nuclear-test-veterans-legal-action-update-23rd-may-2025>, accessed 23 September 2025.

Operation Hurricane, film directed by Ronald Stark, produced by Stuart Legg, UK Ministry of Supply, 1952, <www.iwm.org.uk/collections/item/object/1060022141>, accessed 25 March 2025.

Operation Mosaic (detonations on 16 May 1956 and 19 June 1956 at Monte Bello Island, W.A.), Military Compensation SRCA Manuals and Resources Library, Liability Handbook, Ch 30 Appendices, 4.3 'Operation Mosaic', Australian Government Department of Veterans' Affairs, <clik.dva.gov.au/military-compensation-srca-manuals-and-resources-library/liability-handbook>, accessed 20 January 2025.

Penney, William, interview, 'Atom Man's Hush-Hush Return: Dr. Penney Returns From Monte Bello', British Pathé, 20 October 1952, <www.youtube.com/watch?v=nbRqFhLCic4>, accessed 20 March 2025.

Penney, William, Memorandum to Sir Frederick Brundrett, 22 December 1955, in Jordi Bigues – Archive Palomares, UAB Digital Document Repository (ddd.uab.cat), item PALO_0276, <ddd.uab.cat/pub/palomares/1955/PALO_0276_WGPenney_22des1955.pdf>, accessed 26 September 2025.

Penney, William and Macklen, VHB, obituary, William Richard Joseph Cook, 10 April 1905–16 September 1987, <royalsocietypublishing.org/>, accessed 9 January 2024.

'Phillpot, Henry Robert 1919–2018', *Encyclopaedia of Australian Science and Innovation*, <www.eoas.info/biogs/P003294b.htm>, accessed 29 September 2025.

Radiation Sickness, Medline, <medlineplus.gov/ency/article/000026.htm>, accessed 10 February 2025.

'Radioactive Leeuwin Land Rover "A Pub Tale": Claim', *Fremantle Herald*, 27 November 2015, <heraldonlinejournal.com/2015/11/27/radioactive-leeuwin-land-rover-a-pub-tale-claim/>, accessed 30 September 2025.

'Radioactive Riverside?', *Fremantle Herald*, 20 November 2015, <heraldonlinejournal.com/2015/11/20/radioactive-riverside/?utm_source=chatgpt.com>, accessed 30 September 2025.

Sainsbury, AB, obituary, Vice Admiral Sir Hugh Martell, *The Independent*, 25 February 1999, <www.independent.co.uk/arts-entertainment/obituary-viceadmiral-sir-hugh-martell-1073024>, accessed 14 January 2025.

'The Monte Bello Expedition. A Pictorial Record of the Joint Expedition Undertaken by *The West Australian* and *The Daily News* to Provide a Cover of Britain's first Atomic Explosion – October 1952', BA1716, Sir James Cruthers Collection, JS Battye History Collection, State Library of Western Australia, <slwa.wa.gov.au/pdf/pictorial/BA1716.pdf>, accessed 28 September 2025.

The Plym Pudding (author unknown), newsletter. In <www.labrats.international/post/the-plym-pudding>, accessed 7 March 2025.

This Little Ship, Adrian Cooper and Ron Osborn (directors), UK Atomic Energy Authority, 1953, <www.sonicbomb.com/v1.php?vid=content/atomic/ifilms/this_little_ship&h=360&w=480&ttitle=This%20Little%20Ship%201953&d=>, accessed 20 March 2025.

Trial (1622/05/24), Trial Rocks, Western Australian Shipwreck Databases, Western Australian Museum, <museum.wa.gov.au/maritime-archaeology-db/wrecks/trial>, accessed 9 January 2025.

'Tsar Bomba', Cold War History, Atomic Heritage Foundation, <ahf.nuclearmuseum.org/ahf/history/tsar-bomba>, accessed 13 March 2025.

'UK Atmospheric Nuclear Tests, Factsheet 5: UK Programme', <assets.publishing.service.gov.uk/media/5a755567e5274a59fa717577/ntvfactsheet5.pdf>, accessed 29 September 2025.

ACKNOWLEDGEMENTS

This book is the third in my trilogy on the history of the British atomic tests in Australia. I feel immensely grateful to have had the opportunity to write in depth on these issues. This era of history is of great importance to Australia, and beyond, and to the people who were in various ways affected by these events. In this book, and the others, I have attempted to shed some light on the scientific and political complexities, in an effort to bring the major themes of the test program to a broad audience. As ever, I would like to make it clear that, while many people have contributed to my understanding of the British atomic tests, this book represents my own interpretation of the evidence. Therefore, any errors are mine alone.

I am deeply indebted to Andrew Collett AM, long-time lawyer for Maralinga Tjarutja and counsel to Indigenous people at the Royal Commission in the mid-1980s. In June 2025, in one of his numerous helpful acts, Andrew offered to ferret out files from the National Archives of the UK, in addition to those I obtained during my own trip to London in March 2024. His work in copying and sending those files is greatly appreciated. I would also like to give grateful thanks to Mike Chainey, who efficiently and cheerfully assisted me in fetching and photographing British files in London, in the final stages of writing this book when the pressure was really on.

I owe much to the generosity of Perth-based author Paul Grace, who wrote the excellent book *Operation Hurricane* and has an abiding interest in (and encyclopaedic knowledge of) the events at Monte Bello. Paul shared documents with me during the writing of this book, including a cache of documents from Rory Milne, son of the *West Australian* journalist Norm Milne, who reported the first British test from his vantage point high on a hill on the mainland. Paul was among the group who travelled with me to the Monte Bello Islands in July 2025. In addition to Paul, I travelled with Darren Holden, Toby Holden, Jesse Boylan, Merilyn Fairskye, Maxine Goodwin and Gary Blinco. I simply cannot imagine better travel companions. We shared a unique adventure, bringing our diverse understandings and creative lenses with us. Each night of the trip we made mini-presentations of our own stories and areas of expertise, and had enriching conversations, adding greatly to the sense of discovery we all shared. The islands themselves were other-worldly, and it felt to me like we had stepped out of real life into a spellbinding little archipelagic bubble. Thanks and admiration to the sterling crew of the good ship *Two Seas*, who took us out to the islands. Skipper Steve, cook Ella and deckhand Will are the best in the business, and made the trip both a highly fruitful endeavour and a joy. Not able to join us, but there in spirit, was Honorary Associate Professor Paul Brown from the University of New South Wales, one of the instigators of the trip. Sincere appreciation to Paul for contacting me when he was starting to develop plans.

Many thanks to scientist Madison Williams-Hoffman, who has guided my understanding of recent scientific investigations at the Monte Bello Islands and read a relevant part of the manuscript, as did researcher Dr Mathew P Johansen. Maddy earned her PhD in 2025 for her outstanding research in this area.

The resources company Citic Pacific did excellent work in helping us to identify once and for all the exact location of the lookout on the mainland from which journalists witnessed Operation Hurricane. This was in response to my query to geologist/historian (and dear friend) Darren Holden during 2024, when we were discussing the Monte Bello trip, about whether we could visit the site. Should be easy, right? Just find Mount Potter on a map and head there. Not quite. My enquiry to Darren sparked months of frenetic activity as Citic Pacific picked up the challenge and ran with it. We soon encountered the difficulty that the site was not actually at Mount Potter, as everyone had assumed. So dedicated were the Citic Pacific team, led by Rob Newton, to the venture that they named their efforts Project Hurricane, and got caught up in the excitement. Their hard, diligent groundwork led to the discovery that the lookout was 12 kilometres south of Mount Potter. I had the enormous privilege of climbing to the very place where the journalists camped ahead of Hurricane, up a steep and rocky hill covered in spinifex, ably assisted by Gary Blinco, who took on the role of sherpa but is usually a heritage officer for Citic. We also trekked down to the billabong where the journalists' camp was located. Early on that day, Aboriginal elder Ian Wally, son of Dorrie Wally who witnessed Hurricane, gave a moving Welcome to Country in the cool of the dawn light and granted his permission for me to use the Aboriginal name for the hill the journalists camped on, Bulyroona. I would like to note that there may be several possible spellings of the name, but I have gone with this one.

My sincere thanks to my employer, James Cook University. In particular, I would like to thank my work supervisor, the Dean of Graduate Research and Researcher Development, Professor Stephan Riek, for his support. I also thank the previous acting Dean, Professor Ian Atkinson, who was pivotal in securing my study leave to work on this book. Thanks also to the JCU Research Division and the College of Arts, Society and Education (CASE) for funding my research travel. Special thanks to CASE Dean Professor Koen Stapelbroek, who so eloquently launched my Emu Field book at JCU and has been exceptionally supportive of my research for this book.

I have accessed records from a variety of libraries and archives for this book. In particular I would like to acknowledge the National Archives of the UK, the National Archives of Australia, the State Library of Western Australia Battye Reading Room, and the Australian Institute of Aboriginal and Torres Strait Islander Studies. The JCU Library has also been integral to this work, locating and providing a wide range of documents – some quite obscure – upon request.

I recognise with admiration the work of Susie Boniface, a British journalist who has campaigned vigorously for many years, horrified by the harms done to military personnel and their families by the British tests. Her passionate advocacy has made real inroads into the immoveable object that is the British government. I also salute the children of British veterans, Alan Owen and Steve Purse, who have been leaders in the campaign for recognition of the casualties of the tests while dealing with their own health issues that resulted from their fathers' service. I met these dogged campaigners in London while researching this book, and later joined them in the documentary, *Britain's Nuclear Bomb Test Scandal: Our Stories*, produced by Simon Rawles and screened on BBC TV in November 2024, that showed the terrible aftermath. Also interviewed for that documentary were two Australian campaigners, Karina Lester and Avon Hudson, who have both been stalwart in their work over many years to press for accountability and truth-telling. I am filled with admiration for what they have achieved in their quest.

While I was visiting the Monte Bello Islands in July 2025, my dear Dad, Frank, died. The momentous loss of my Dad, while I was actually on the ground at MBI, makes the trip even more poignant and significant for me. I am grateful to Dad, and to the rest of my family, for insisting that I proceed with my trip, despite his severe health issues in the months leading up to my visit. Love and thanks to my family – Mum Rosemary, Stepmum Inta, Meredith, Andrew, Narelle, Sophie, Alex and William. As always, my dear friends and my animal family have all had a part as well, keeping me grounded and cognisant of the need to stop work from time to time. Thanks also to my wonderful neighbours at Alligator Creek, who are especially appreciated when the cyclones come spinning through our community. Shout out to my trainer Neff at Plus Fitness, who has played an important role in making me strong in recent years.

The team at NewSouth are always so great to work with, and this is the third time around so I speak from considerable experience. Thanks to exceptional publisher Elspeth Menzies, who is always positive, supportive and perceptive. Project editor Joumana Awad has been outstanding, the calm centre of the whirlwind. John Mapps has edited with much-appreciated skill and sensitivity. Thank you to proofreader Gabriella Sterio and indexer Jenny Browne for their detailed work.

The story of the British nuclear tests in Australia is far from complete. While many of us are providing more and more pieces of this vast jigsaw, parts of the picture are still distressingly blank. Despite some document releases (and re-releases) recently, many British files remain secret, kept from the eyes of historians, survivors and descendants. This is unacceptable. Australians have a right to know exactly what took place on our soil. I would like the British government to finally make available all relevant information about the tests. In making this plea I do not have any expectation of a positive response, but it is incumbent on everyone who delves into this murky story to add their voice to calls for accountability.

INDEX

THE SECRET OF EMU FIELD

BRITAIN'S FORGOTTEN ATOMIC TESTS IN AUSTRALIA

Elizabeth Tynan reveals the story of a cataclysmic collision between an ancient Aboriginal land and the post-war Britain of Winston Churchill and his gung-ho scientific advisor Frederick Lindemann. The presence of local Aṉangu people did not interfere with Churchill's geopolitical aims and they are still paying the price. The British undertook Operation Totem at Emu Field under cover of extreme remoteness and secrecy, a shroud of mystery that continues to this day.

Scan to read more

Printed and bound by CPI Group (UK) Ltd, Croydon, CR0 4YY

16/09/2026

14963560-0005